Technology and Resistance

Studies in the Postmodern Theory of Education

Joe L. Kincheloe and Shirley R. Steinberg
General Editors

Vol. 59

PETER LANG
New York • Washington, D.C./Baltimore • Boston • Bern
Frankfurt am Main • Berlin • Brussels • Vienna • Oxford

Technology and Resistance

Digital Communications and New Coalitions around the World

HM
221
.T298
2000
West

Edited by
Ann De Vaney,
Stephen Gance,
and Yan Ma

PETER LANG
New York • Washington, D.C./Baltimore • Boston • Bern
Frankfurt am Main • Berlin • Brussels • Vienna • Oxford

Library of Congress Cataloging-in-Publication Data

Technology and resistance: digital communications and new coalitions
around the world / edited by Ann De Vaney, Stephen Gance, and Yan Ma.
p. cm. — (Counterpoints; 59)
Includes bibliographical references and index.
1. Technology—Social aspects. 2. Technology—Political aspects.
3. Information technology—Social aspects. 4. Information technology—Political
aspects. 5. Information society—Political aspects. I. De Vaney, Ann. II. Gance,
Stephen. III. Ma, Yan. IV. Counterpoints (New York, N.Y.); vol. 59.
HM221.T398   306.4'6—dc21   97-017112
ISBN 0-8204-3795-6
ISSN 1058-1634

Die Deutsche Bibliothek-CIP-Einheitsaufnahme

Technology and resistance: digital communications and new coalitions
around the world / ed. by Ann De Vaney, Stephen Gance, and Yan Ma.
–New York; Washington, D.C./Baltimore; Boston; Bern;
Frankfurt am Main; Berlin; Brussels; Vienna; Oxford: Lang.
(Counterpoints; Vol. 59)
ISBN 0-8204-3795-6

Cover design by Nona Reuter

The paper in this book meets the guidelines for permanence and durability
of the Committee on Production Guidelines for Book Longevity
of the Council of Library Resources.

© 2000 Peter Lang Publishing, Inc., New York

All rights reserved.
Reprint or reproduction, even partially, in all forms such as microfilm,
xerography, microfiche, microcard, and offset strictly prohibited.

Printed in the United States of America

Dedication

We dedicate this book to the numerous international students who have graduated from our program in Educational Communications and Technology at the University of Wisconsin-Madison and returned to their countries to be leaders in the equitable use of technology in educational and political arenas.

Acknowledgment

We would like to thank Mary Jane Curry for her invaluable help in designing and editing this book, and Sousan Arafeh for her assistance in writing and compiling the proposal for this book.

The Free Burma Poster is reprinted with the kind permission of David McLimans, Madison, WI, 53706.

The map of Africa and Malawi is reproduced from the Perry-Castañeda Library Map Collection available at www.lib.utexas.edu/Libs/PCL/Map_collection/Map_collection.html

Correspondence should be addressed to:

The General Libraries
PCL Map Collection
The University of Texas at Austin
Post Office Box P
Austin, Texas 78713-891
PHONE: (512) 495-4350
FAX: (512) 495-4347

Table of Contents

Introduction   1
Ann De Vaney
Stephen Gance

Chapter One   9
Technology in Old Democratic Discourses and Current Resistance Narrative: What is Borrowed? What is Abandoned? What is New?
Ann De Vaney

Chapter Two   51
Breaking the Silence: Fax Transmissions and the Movement for Democracy in Malawi
Kedmon N. Hungwe

Chapter Three   71
Resistance and Cybercommunities: The Internet and the Free Burma Movement
Zarni

Chapter Four   89
Old Technology in New Contexts: Print Media and Russian Education
Stephen T. Kerr

Chapter Five   113
Women, Telephones, and Subtle Solidarity: A Counternarrative
Sousan Arafeh

Chapter Six 139
Chinese Online Presence: Tiananmen Square and Beyond
  Yan Ma

Chapter Seven 153
Computer Links to the West: Experiences from Turkey
  Marina Stock McIsaac
  Petek Askar
  Buket Akkoyunlu

List of Contributors 167

Index 171

Introduction

Ann De Vaney and Stephen Gance

> As rebellions broke out across Indonesia this month, protesters did not have tanks or guns. But they had a powerful tool that wasn't available during the country's previous uprisings: the Internet. . . . Bypassing the government–controlled television and radio stations, the dissidents shared information about protests by e-mail, inundated news groups with stories of President Suharto's corruption, and used chat groups to exchange tips about resisting troops. In a country made up of thousands of islands, where phone calls are expensive, the electronic messages reached key organizers.
>
> *Boston Globe,* 23 May 1998

This book is a political project providing a platform for voices and narratives of heretofore powerless resisters who use decentralized technologies to advance their democratic causes. We support them in their efforts to capture some of the flow of information within their various countries. This is not a book about the politics of nongovernmental organizations, nor about the current tension between nation and state. It is the first step in a political undertaking that offers description of dissident activity, rhetoric, and the use of technology.

2

Decentralizing Information

The proliferation of digital communications and the growing presence of portable computers have placed the means of production and dissemination of news and messages in the hands of almost any political group. Resistance movements in South Africa, Namibia, Malawi, China, Russia, Indonesia, and Mexico have relied on ready access to communication technologies for promulgation of their liberatory messages. Citizens who enjoy democratic rights have joined in solidarity with members of oppressed nongovernmental organizations (NGOs) through the use of contemporary links such as electronic mail and facsimile. Some would say that the geography of cyberspace is the new arena for conversation about political resistance. While these are hopeful events, they often take place under the gaze of oppressive governmental agencies who may yet move to redirect the flow of information away from grassroots movements, just as official Chinese censors did after the Tiananmen Square protest; that event prompted them to read, temporarily, all email messages leaving their country. The possibility that new discourses in cyberspace could address and aid in the redress of disenfranchised citizens in many countries must therefore be tempered by the political realities of ruling governments. It is the dominant state agencies and corporate organizations who still control the major flow of information around the world.

Disrupting the State

Decentralized electronic technologies have already caught the governments of Russia, Mexico, and China off guard. While we would not go as far as Castells (1998) when he writes that these governments must now share power with grassroots organizations, we would agree that they have disrupted the flow of publicity and news that their governments promote. And, such actions have lifted the mask from an element of the state's legitimating function, have thwarted some of its power to elicit normative behavior from citizens, and have hampered its ability to circulate norms for patriotic behavior. The new information technologies have contributed to a decentralization and an undermining of bureaucratic or governmental control (Castells 1997). These decentralizing and democratizing tendencies have largely been at the expense of the state and in spite of the vast informational and technological apparatus that the state controls. Indeed, as can be seen in the case of Burma and China (and now Indonesia), information dissemination separate from the state was accomplished using the technological infrastructure created by the state.

This is not to claim that information technologies are always liberatory or democratic. Certainly, the state can still muster considerable influence over the dissemination of information as indicated, for example, in China, where the Chinese government has been effective in controlling the flow of information in the years after the Tiananmen Square crackdown. Castells (1997) goes so far as to claim that oppressive societies may become more oppressive with the new information technologies, whereas open societies may become more open. We do not wish to make such a claim. Rather, the chapters here represent examples where information technologies have been effectively applied in liberatory projects, but these technologies do not guarantee democracy or democratic participation in any society, oppressive or not.

Whose Interests Are Met?

Newspapers and Web sites provide timely reports of the events and strategies of resistance movements around the world. In May 1998 there was news of how Indonesian activists used the Internet to skirt press censorship and rouse citizens to action. Dissidents circulated stories of the 1986 "people's power" rebellion on the Internet to remote citizens. When students occupied the Indonesian parliament building, one leader carried in a laptop computer and went online while surrounded by troops. Similarly, pleas for help and recognition of the plight of the Indians in Chiapas, central Mexico, have gone out in numerous languages on the Internet.

It is important to note that the success of the dissemination of political publicity in Marxist or democratic resistance efforts is at least partly the effect of the power of the flow of information. We need only recall that Mussolini is credited with being the first modern ruler to control radio waves for the promulgation of his fascist message. Nor can we forget that probably the most effective use of mass cinema was under the direction of Goebels in the 1930s for the purpose of turning non-Jewish German citizens against Jewish German citizens. It is from the demagogues of the twentieth century that dominant and subversive groups have learned about the power of the flow of information. What differs in the resistances described here is that dissidents are working with demassed, personal technologies.

Underestimating citizens. Dominant political processes in the West are increasingly dependent on media for their actualization (Castells 1997). Indeed, Castells claims that the political use of the media is a necessary (albeit insufficient) requirement for successful political mobilization. It should be no surprise, then, that groups seeking the overthrow of repressive regimes would turn to the media to disseminate their message. Castells reminds us of the basic propaganda model of media which makes the false assumption that media has the unconditional power to persuade a mostly passive audience. Centralized media within countries ruled by oppressive regimes operates, to some degree, on this principle. There seems to be a certain belief in the efficacy of the media to maintain control. The Internet and other electronic communications then become sites of uncontrolled space and thus spaces of political struggle (de Certeau 1984). Within countries with oppressive regimes (or centralized, bureaucratic regimes, e.g., Russia) information technologies offer one of the few sites for democratic action. But, as de Certeau notes, these spaces will always be shifting as governments and groups vie for their control. Thus, the stories here represent a particular moment in history where information technologies have been utilized for democratic ends. No doubt, as in all politics, spaces of resistance will open while others will close.

Whose voices are heard? The Internet, email, and faxing are still relatively accessible (as compared to television, for instance) for groups with few resources. Yet these technologies, by their very nature, can be a barrier to democratic participation. One might question, for instance, the ability of *campesinos* in Mexico, who cannot read and who have no electrical power nor phone lines, to access computers and the Internet. Even in the United States, access to the Internet is not equal across all social groups, a point made by both De Vaney and Arafeh, thus undermining the democratic rhetoric of

Internet proponents. Further, once liberatory movements succeed in securing political change, the processes of democracy are arduous and depend upon a recognition of the multiple ethnic and political groupings within the country and thus require a pluralistic democratic process. As Chapter One makes clear, such is the case in the fledgling democracies of South Africa and Namibia, where attenuated discourses of democracy present in the struggles for freedom that preceded political change have become more fully articulated as democratic processes continue. Technology has the potential to play a different role in developing nations, often becoming a source of exclusion rather than liberation because of the technical skills and equipment required to participate in a technology-based democratic forum. But the chapters here relate stories of initial efforts, of fledgling democratic strategies.

Resistance Narratives

Common themes emerge in the telling of the technology and resistance stories in this text. Each chapter describes a separate reconstruction of a site of power and the establishment of new discursive space for purposes of resistance and liberation. Such discursive formations are not entirely new. They formerly existed around small printing presses and some cable television applications. However, there are new discourses coalescing around current personal digital technologies. Inasmuch as the discursive space surrounding these technologies had already been established, it has worked to regulate or, in some cases, proscribe usage. Paul Bove (1992) writes that discourse acts as power's relay throughout a social system, and in each of the accounts related here, the sites of power around a technology had to be restructured. Such restructuring is hard in any social or political setting, even if one lives in a democracy. Western rhetoric about these technologies spreads with their proliferation throughout the world, and utopian and dystopian narratives about electronic communications had to be countered by coalitions working against the grain, appropriating these technologies for their liberatory projects.

Overview of the Book

In Chapter One, De Vaney frames the various essays in this book by questioning the juxtaposition of technology and democracy within these political projects. She points out that all of the coalitions represented in this text are creating new discursive space using rhetoric that inevitably draws on discourses of democracy spanning several centuries and containing liberal as well as Marxist threads. Highlighting the new subjectivities created by these virtual spaces, De Vaney describes how these digital technologies realign compatriots while structuring new meanings and uses.

In "Breaking the Silence: Fax Transmissions and the Movement for Democracy in Malawi," Hungwe describes the harsh suppression of democracy under Banda, the Malawian president for life, and the emergence of a democracy movement which ended that rule. Fax transmissions proved to be an effective means for the emergent democracy movement to secure alternative ways of disseminating information independent of the state-controlled media. Hungwe recounts the use of faxing within the context of the broader changes that took place in Malawi in 1992.

Discourses of democracy present on and about the Internet (see Chapter One) contribute to its legitimacy as a credible source of news about various resistance movements. Thus, the dissemination of resistance information via the Internet will inevitably contribute to and change the discursive space of democracy and technology and take on symbolic meanings circulating within that space. In this way, the politics of resistance takes on a global character that is in some sense independent of its internal effects to the country that is the target of the resistance. Burma, China, and to a lesser extent, Malawi are cases in point where international pressure on corrupt or authoritarian regimes is considered a key part of political resistance.

In "Resistance and Cybercommunities: The Internet and the Free Burma Movement," Zarni discusses the internal communicative culture of the Free Burma movement in which the Internet plays an essential role. People from nineteen different countries have used the Internet to form coalitions and share strategies in their efforts to weaken the grip of the military rulers of Burma. Specifically, they have taken advantage of the Internet to develop communities of understanding, to educate a global audience about the political situation in Burma, and to implement actions against multinational corporations who are investing in Burma. For Burmese dissidents in particular, the Internet has provided a voice for people who are silenced in their own country because of the control of the media by the military junta within the country.

In the next two chapters, the emphasis of the use of technology in the support of coalitions is not on overthrowing oppressive or authoritarian governments (although, in the case of Russia, these elements certainly existed). Rather, the emphasis is on the appropriation of technological resources by marginally franchised people within these countries. These chapters highlight the various coalitions and solidarities that arise around technologies which provide a focal point for organizing. In "Old Technology in New Contexts: Print Media and Russian Education," Kerr discusses print as a revolutionary medium for Russian schools. Noting the Russian respect for and mystical attachment to the printed word as a vehicle for revolutionary change, Kerr details the slow but persistent effort to improve pedagogy in the educational system by organizing and educating teachers and by bringing reliable instructional materials to students. He reminds us that books, journals, and newspapers, seen as old-fashioned in the West, are still powerful tools for most Russian teachers.

In "Women, Telephones and Subtle Solidarity: A Counter-narrative," Arafeh describes another neglected, personal "information technology," the telephone. She celebrates the telephone's contradictory but positive effect on the lives of women. The telephone, she notes, remains the primary communications technology in the United States, especially for low-income people, many of whom are women, racial and ethnic "minorities," and/or single heads of households. Arafeh argues that in our haste to embrace the latest technological obsession, we should not forget foundational communications technologies like the phone and its role in quietly doing the lion's share of maintaining ongoing, essential connections and coalitions, particularly for women.

The last two chapters explore the use of decentralized technology for purposes of dissident action in China and formation of global academic coalitions in Turkey. The use of the Internet and fax during the Tiananmen Square protests in China had an important internal as well as a global effect. In "Chinese Online Presence: Tiananmen Square and Beyond," Yan Ma recounts the use of fax for disseminating information during the Tiananmen Square incident in the spring of 1989. Chinese students used the occasion of the memorial service for the late Hu Yaobang, a reformer among the top Chinese Communist leaders, to gather in Tiananmen Square. The students' demands were focused on free expression and an end to the corruption of the Communist Party in China. The government leaders refused to meet the demands of the students. Later, when hard-liners in the government gained control of the media and shut down satellite transmissions, faxed messages became an important source of information about the protest, both inside and outside China.

In "Computer Links to the West: Experiences from a Turkish University," McIsaac, Askar, and Akkoyunlu report on projects which introduced the use of electronic mail to certain faculty and students in various academic departments at Turkish universities from 1986 to 1992. Although the initial project was designed simply to introduce a new technology, the subsequent utilization of computer-mediated communication had social, political, and cultural implications far beyond the university context.

References

Bove, Paul. 1992. *Mastering Discourse: The politics of intellectual culture.* Durham, NC: Duke University Press.
Castells, Manuel. 1998. *End of Millennium.* Malden, MA: Blackwell Publishers.
———. 1997. *The Power of Identity.* Malden, MA: Blackwell Publishers.
de Certeau, Michel. 1984. *The practice of everyday life.* Berkeley: University of California Press.
Marcus, David L. 1998. Indonesia revolt was Net driven. *Boston Globe Online,* 23 May, 1998. (http://www.boston.com/dailyglobe/globehtml/143/Indonesia revolt was Net driven.htm)

Chapter One

Technology in Old Democratic Discourses and Current Resistance Narratives: What Is Borrowed? What Is Abandoned? What Is New?

Ann De Vaney

Since this book is the start of a political project to uncover stories of resisters using decentralized technology, I will explore the space of their narratives. These chapters, by their very rhetoric, investigate the relationship between certain forms of technology and emergent or resistive democratic movements. The appropriation of communication technologies by citizens in these movements with little prior political power will most likely result in new discursive formations within accounts of their practices. In such cases seemingly disparate technological and democratic discourses might blend in the new discursive space. Certainly new subjectivities will emerge in rhetoric used to express old democratic concepts.

Since one can only know about these revolutionary or rebellious events through written or oral accounts, a listener's understandings are mediated by language. Questions about discursive formations around these new sites of power are central to comprehension. To make sense, therefore, of new discursive spaces created by these global freedom fighters and coalition participants who employ technology, I will explore the manner in which the concept of technology is related to that of democracy in old and new discourses. Noting the manner in which democracy has become a diffuse sign, I will discuss the relation of specific resistance movements to this dispersion. In an attempt to trace the values and attitudes expressed in the resistance discourses, I will describe what I consider the formation of a modern technology discourse as it unfolded within democratic texts in the nineteenth century and update the discussion for the twentieth century. Finally, I will explore the subjectivities of these new democratic discourses and summarize my claims about the global freedom fighters, coalition participants, and resistance movements.

I take my understanding of the action of democratic movements from author accounts, and my use of terms such as democracy, capitalism, markets, technology, progress, communication technology, modernity, postmodernity, and the New World Order from the historical and contemporary rhetorical constructions in which they are embedded. In other words, my understanding of democracy from the pen of Jefferson comes from a partial Rousseauean discourse in which subjects are romanticized. The point is this: Meanings in this essay are suggested by the discourses that construct, produce, and frame those meanings. Since the concepts that surround the issues in this book are multifaceted and fluid, such a discursive approach will provide more explanatory power than one in which terms such as democracy and technology are considered unified concepts with fixed definitions.

Overblown Claims

Claims have been made about the Internet as a medium for the promotion of participatory democracy on a national level, but the accounts here are not about these overblown claims. I remain skeptical about the articulation of these rhetorical assertions. How, for example, could the Internet promote participatory democracy in the United States, when computers remain in the hands of the professional classes? Lower income groups, in which people of color are over-represented, would become more disenfranchised in an

Internet-based participatory democracy than they are today. Likewise, I suspect, political parties and their leaders would be loathe to give up control of the flow of certain political information. For instance, many candidates running for office in the late nineties established Web sites to promote themselves and their political agendas, but used the electronic address as a one-way channel; they posted selected information about themselves and their issues on the sites.

In this chapter I am making a smaller claim. I notice a trend within attenuated democratic and resistance movements to appropriate information technologies for the promulgation of messages and advancement of causes. Yet, communication technology is generally controlled by a few corporate giants or by a centralized state bureau. One might speculate that aspects of the power to control the flow of information have been breached and centralized control has been challenged. Castells (1998) believes that the power of small coalitions, sometimes nationalist or ethnic, have become co-existent with centralized state powers and that these fractured entities share governance. But one can see that elsewhere, for example, in China, other independence movements employing technology have failed and their right to transmit information has been abrogated. Both problematic outcomes help define our postmodern world[1] and deserve exploration.

Skirting the Modern and Postmodern

The stories of revolutionary or resistive events here and elsewhere are recounted in a modern voice; readers hear liberal ideas and the values of democratic discourses unfolding in the tales. But the overthrow of contemporary authoritarian governments follows hard on the heels of the overthrow of literary and artistic canons. These events of resistance take place in a postmodern world where nation as well as image is fractured, where identity and subjectivity are fragmented, where narratives are nonsequential and messages defy expository rules, where space is real and virtual, and where democracy, some think, can be reduced to keywords (e.g., Williams 1976). One author believes that currently there is disorganized capitalism, "disjunctures between and among the economy, culture and politics" (Appaduri 1990, 296).

The resurgence of interest in democratic politics, however, is taking place not only in the sites described here, but in the academy as well. Many postmodern scholars who deconstruct sites of power in the name of disenfranchised peoples have pragmatic projects. Their agendas are bifurcated; they seek modern remedies such as the installation of egalitarian principles, but they often use postmodern techniques such as deconstruction to locate inequitable sites of power. According to Herrnstein Smith (1988), some postmodern projects have allowed traditional disciplines to expand the acceptable objects of study, modes of operation, and audiences. By describing new projects for technology and democracy, the authors here reconstruct or reinscribe postmodern sites of power. Carrying the weighty modern signs of technology and democracy, they describe the intersection of radical forms of democracy with appropriated electronic communications. They do this in a world where democracy has broad powers and discursive space; however, many liberal concepts have been vacated from these democratic discourses.

New Discourses and Old Rhetoric

Political projects that combine the power of democratic concepts with the use of technology have existed since the time of the Enlightenment and, strangely enough, Enlightenment rhetoric is apparent in the resistance stories in this text and narratives from other dissidents. To understand the contemporary freedom fighters, it is helpful to explore the historical scource of their rhetoric. Since technology has been so implicated in the articulation of those Enlightenment goals which eventually oppressed workers instead of freeing them, some critics tend to fault democratization itself. During the industrial revolution the factory system helped create a managerial class and fostered the growth of a bureaucratic (business and economic) hierarchy while automation created a class of deskilled laborers. With a critical focus on these and other social ills, it is hard for us to remember the initial connection of technology to democratic freedoms that early Enlightenment writers espoused.

The relationship of technology to democratic movements in the modern world has been double-edged and always conflated with capitalism. Hailing the beginning of the American Enlightenment and modernity in the New World, Jefferson and Franklin described the potential of technology to foster democratic activity. In this view, citizens would be partially freed from drudgery and could participate more fully in the body politic. Further,

> Benjamin Franklin and Thomas Jefferson ... were true believers in humankind's steady moral and material improvement. As avid proponents of the cause of liberty, they looked to the new mechanical technologies of the era as a means of achieving the virtuous and prosperous republican society that they associated with the goals of the American Revolution. (Marx and Smith 1994, 3)

Yet their belief was tempered.

> Jefferson worried the civilizing process of large scale technology and industrialization might easily be corrupted and bring down the moral and political economy he and his contemporaries had worked so hard to erect. As much as Jefferson esteemed discovery and invention, he considered them means to achieving a larger social end. For his part Benjamin Franklin refused to patent his inventions ...[he] considered his inventions not a source of private wealth but a benefit for all members of society. (Marx and Smith 1994, 3)

Jefferson and Franklin invoked technology to help build a healthy republic, but in doing so their rhetoric constructed a romantic subject. Hamilton, on the other hand, had a less humanistic and more technocratic project. He saw technology as fostering economic growth which would in turn maintain a fiscally strong republic.

> Evident in the speeches and writings of Alexander Hamilton and his associate at the U.S. Treasury Department, Tench Coxe, was this new viewpoint that attributed agency and value to the age's impressive mechanical technologies and began to project them as an independent force in society. (Marx and Smith 1994, 4)

Hamilton considered manufacturing under the factory system as political salvation, as freedom from England and Europe, and envisioned this system as one that would maintain political order.

These divergent American views were borrowed from British, Scottish, and European Enlightenment scholars (Ferguson 1994), but they contain the seeds of most modern discussions about the manner in which technology relates to efforts to form a democratic society. Even though Franklin and Jefferson focus on the improvement of the individual, they attribute social powers to technology that machines simply do not possess. While sacrificing the individual to the body politic, Hamilton attributes political powers to the factory system. Both approaches are deterministic, one romantic and one technocratic. This deterministic thread has haunted popular and scholarly discussions of technology from the time of the Enlightenment until today. Furthermore, arguments from the proponents of technology in democratic movements likewise fall into the two camps described here. Competing but complementary themes about this relationship may be expressed as 1) technology provides individuals with the freedom to become better citizens, or 2) technology provides a stable political economy. These separate deterministic arguments are inscribed in the writings of eighteenth- and nineteenth-century scholars in whose texts the overarching concepts of democracy, technology, and capitalism, especially the market economy, become hopelessly conflated. Each of these concepts, however, achieves the position of a central sign in the discourses of modernity. The centrality and importance of these signs disperse their significations, allowing a surplus of meaning to spill over into signifiers within many divergent discourses. In modernity these signs are everywhere, but nowhere is it easy to identify their original significations.

Democratic Discourses

Various incarnations of democracy have spawned diverse models of that form of government. While espousing the "rights of man," certain freedoms, equality, and self-rule, most Western revolutionaries of the eighteenth and nineteenth centuries sought to establish fully articulated democracies. These included liberty, equality, and self-rule; the guarantees of certain freedoms, including freedom of speech, assembly, and the press; and the guarantees of certain rights, including voting, owning private property, and bearing arms. Rich democratic discourses articulating multifaceted democratic ideas and liberal values emerged before, during, and after the founding of these new governments and contributed to the project of modernity. In this postmodern world democracy is still valorized.

> We are witnessing the return of the idea of human rights. That idea is stronger now than ever because it has been sustained by resistance fighters, dissidents, and critical thinkers who fought totalitarian powers in the darkest hours of our century. From the workers and intellectuals at Gdansk to those in Tianannemen Square, from U.S. civil rights activists to the European students of May 1968, from those who fought South African apartheid to those who are still fighting dictatorship in Burma, from Chile's Vicaria de la Solidaridad to the Serbian opposition, from Salman Rushdie to embattled Algerian intellectuals, the democratic spirit has been brought to life by all who have opposed increasingly absolute powers in the name of their right to live in freedom. (Touraine 1997, 10)

Revolutions and Democracy

Democracy as a modern sign is diffuse and occupies broad discursive space, but that space may be constrained in the postmodern world. In his notice of freedom movements around the world, Touraine (1997) indicates that they may have only one practice in common, the overthrow of a dictator or oppressive government. If original democratic discourses included a nexus of liberal concepts such as self-rule, guarantees of certain freedoms and equality as well as liberty, the focus on the overthrow leaves much of that space vacated and occupied by fewer ideas. If the democratic space has been attenuated and fewer freedoms exist, except the will of the people to overthrow and their desire to free themselves of specific oppressions, then the smaller space is weakened and subject to attack. Touraine notes that certain new states, for example, that strive for democracy have no means of handling national groups, such as religious or ethnic coalitions. Another example of what I call an attenuated democracy is that described in Malawi. From Hungwe's report in this text (Chapter Two) I deduce that freedom of speech may be limited by the new regime, since some newspaper editorials were censored by that government.

Since the means of achieving a democratic society have traditionally been revolutions, one cannot discuss democratic discourses without including revolutionary practices. Events related in this text and other global resistance movements fall into two categories: movements that seek to overthrow governments and install new ones (e.g., Russia, South Africa, Namibia, perhaps Malawi, and some countries in the former Soviet bloc); and those that work to gain specific freedoms or guarantees of equitable political and social practices from existing governments (e.g., Indonesia, China, Burma and Mexico).

Because of the social inequities that were outcomes of certain democratic revolutions some critics warn against their effects. Alaine Touraine, who writes about resistance movements, leads the critics of large revolutions that seek to overthrow governments in the name of democracy.

> Revolutions often intended to save democracy from its enemies have instead given birth to anti-revolutionary regimes due to their concentration of power, evocation of national unity and unanimity and denunciation of adversaries, who are deemed traitors with whom it is impossible to co-exist, rather than spokesmen for different ideas or interests. (Touraine 1990, 14)

His quote is an evocation of the chaos in Yugoslavia today, but offers little hope for the well-planned and tempered revolutions of South Africa and Namibia. It is too soon to tell if democratic discourse in these countries will remain comprehensive or, if by virtue of the social, political, and economic difficulty of sustaining such an endeavor, will be ratcheted down.

South Africa and Namibia. Electronic communication links and desktop computers were the technological modes that enabled Africans in their fight to liberate South Africa and Namibia. The way technology unfolds in their stories is important enough to my claims to relate here.

In the late 1980s, Technica, an international organization dedicated to the teaching of technical skills to liberation groups; the African National Congress (ANC); and the South West African People's Organization (SWAPO), a Marxist liberation group, joined forces with Namibian freedom fighters to place specific technologies in the hands of the disenfranchised black majorities. Their efforts helped prepare citizens in these neighboring countries, South Africa and Namibia, for the final elections that placed them in power in the early nineties. Some of the tactics for placement of computers in villages and education of village citizens were similar in both countries. Technica and ANC, or Technica and SWAPO members placed Macintoshes in South African village halls, teaching people to use word processing software and produce newsletters on Adobe Pagemaker. Citizens were then able to distribute their own information about liberation movements and upcoming elections.

In Lukasa, South Africa, Technica members helped change the crucial ANC news outlet, *Rixaca*, to a desktop publication. Prior to that the Department of Information writers who had produced Rixaca mailed it to Europe for printing. The desktop substitute increased the speed of news dissemination, just as resistance to ANC efforts were increasing. While the situation in Namibia was somewhat similar, there were differences that called for separate strategies (Porteus and Pratt 1988).

Technica helped SWAPO establish a desktop-published news outlet, *Namibia Today*, in Windhoek, Namibia, but also disseminated breaking political news by fax to the northern sections of Namibia. The largest contingent of SWAPO freedom fighters lived in the north where the white government had invested little money in roads or phone lines. Poor communication links kept many Namibians ignorant of government activities. Technica and SWAPO linked many sites in the north to Windhoek by phone for the delivery of faxes. One of the ironies of this accomplishment was that the phones were wind-up machines. In addition to maintaining the wind-up connection, operators had to learn not to interrupt calls and to place them immediately for the transmission of a fax. Since the literacy rate was low in northern Namibia, these faxes were composed of graphics as well as text and contained news about how the government was harassing the majority population as elections for independence approached (Porteus and Pratt 1988).

Concern within these revolutions was focused on fairly traditional eighteenth-century democratic goals, and as I listen[2] to liberation stories unfold, technology hangs in the background. The oral and written rhetoric, particularly of the South African movement, is classically democratic with one exception. Eighteenth-century democratic texts written by Western white men assume a universality for their concepts of equality, but guarantee equality only for themselves (I shall return to this topic of race later). That mistake, of course, is not made in the South African liberation texts, but otherwise the rhetoric of the Government of National Unity's (GNU) constitution is evocative of early democratic treatises. The Preamble to the GNU Constitution opens

> We the people of South Africa,
> Recognize the injustices of our past;

And continues

> Lay the foundation for a democratic and open society in which government is based on the will of the people and every citizen is equally protected by the law. (GNU 1996)

In this preamble and in the GNU Constitution, democracy is not attenuated but rather fully articulated as in older discourses, but happily includes the concept of "diversity." While democratic rhetoric, however, continues to hold sway in the functioning of the GNU, technology remains an important subtext. South Africa has the highest number of Internet hosts of any non-OECD (Organization for Economic Cooperation and Development) country (Castells 1998).

Grassroots Alliances

South Africa is not the only geographic site in which resisters or former resisters continue to strengthen their ties to global allies by electronic networking. One scholar (Cleaver 1997) mentions grassroots groups who take advantage of the Internet in Germany, Canada, Italy, and the United States. Describing these nongovernmental organizations (NGOs), he claims that the Zapatista movement in Mexico started the trend of Internet use by grassroots movements in 1994, and calls this trend the Zapatista effect. Castells (1997), while recognizing the importance of the Zapatistas and their challenge to the Mexican centers of government, identifies them as one important movement among many such organizations. And I believe that online organizations in the United States, such as militia groups or environmental coalitions, were established before the Zapatistas. Around the world, electronic alliances sue for citizen "rights" from the Left and the Right; the Zapatistas on the Left and the militias on the Right fight for more local or regional control in the face of state and global power structures (United Nations, International Monetary Fund, World Bank, North American Free Trade Agreement). But libertarian militias, I believe, are actually different from many other Right or Left NGOs. Although, like the latter, they do stand against the New World Order, their desires go beyond what we, in the United States, call a return to state control or "statism." Along with certain conservative lawmakers in the United States Congress, they maintain an interest in a devolution to return control of government to the states, but their actions uncover a stance even against the geographic states in which they reside, so fervent is their libertarianism. They maintain an anarchical position against any form of government. Although anarchy is a democratic tactic of last resort, most global grassroots NGOs assert their desire for democracy in various but less inflammatory ways. (Such is the contemporary nature of that term, democracy, whose referents are dispersed in a Derridean sense, rather than pinpointed in a positivistic sense.) Most NGOs wish to share power with their governments in new, "more democratic" ways, but some encounter insurmountable tensions between the concepts of nation and state. The Zapatistas, for example, when invited to join one of the Mexican political parties, rejected the offer to be part of the existing structure that was causing the problem (Castells 1997, Cleaver 1997). They sought local control, in this case indigenous control, of social and political decisions about the Chiapas region. But is this democratic?

The Zapatistas, alleging democratic goals, are using tactics established for decades in Marxist guerilla movements. In fact, the Marxist existence in Latin America and South Africa was overt with the presence of SWAPO and Technica. Is Marxism in such disarray and disrepute that the dissidents, who use its tactics and have the Marxist goal of overcoming oppression, are reluctant to name it? Is their appropriation of the term "democracy" equivalent to the appropriation of the term "empowerment" by the Right in the United States, or are they truly seeking democratic goals? Has the term "democracy" become a keyword, as Raymond Williams (1976) claimed? And would these dissidents be able to garner individual allies around the world if they packaged their electronic message in a communist box? Do they use the discourse of the dominant governments to open a conversation with citizens of other democratic countries, or do they actually share some of the underlying liberal values of citizens living within democratic governments?

Unfortunately, I am far from being able to answer these questions, but what I read and hear in the tales of resisters in this text and elsewhere is conflated. It is not surprising that the rhetoric of early democratic texts appears in new discourses that claim to be democratic, nor is it surprising that a Marxist search for justice is encompassed in their desires. The original Marxist subject, however, as constructed in early Marxist texts is absent in the stories I hear and an expanded democratic subject appears. Yet an understanding of the social and political power of the means of communication comes originally from Marx and was articulated in all the Marxist guerilla movements of the twentieth century. In the historical section of this chapter I shall try to tease out the discursive antecedents of the rhetoric in these tales of resistance by examining early democratic texts and their relation to technology, and by highlighting Marx's understanding of the power of communication technologies.

Markets

While social justice can be sought under the rubric of either democracy or Marxism in a similar manner, economic justice has a different trajectory in these disparate political discourses. The resurgence of interest in democracy is directly related to the vast economic success of a market economies. Indeed, most critics and pundits say that democracy persists because of its continued pairing with a market economy which has provided many countries with extensive capital growth. Early in the recent Soviet revolution, "market democracy" was the postmodern rhetoric used to describe that governmental experiment, although "crony capitalism" erased that new phrase.

> Many now believe that democracy is of necessity the normal form of political organization, the political face of modernity, whose economic form is the market economy and whose cultural expression is secularization. (Touraine 1997, 7)

An open market is a goal for many resistance movements and although market issues are not directly addressed in the chapters here, other NGO accounts (e.g., South Africa, Namibia, Russia, Mexico) include this goal. Touraine (1997) wants to retain the distinction between markets and democracy; he asserts that a competitive market does not constitute a democracy any more than a market economy constitutes an industrial society. An open market, he notes, is a necessary but not a sufficient precondition for democracy.

China's Markets

China's leaders are currently mounting claims about the adoption of open markets, and their efforts are touted by some United States citizens, especially those with corporate interests who wish to see the federal government continue favored-nation trading status for that country. Some even suggest that the Chinese economy is nearing the conditions for open markets and hence democracy. Yet the occurrences in Hong Kong since the return of the colony to the mainland government belies the ability of an authoritarian regime to establish an open market. "For open markets there must be a legal system, a public administration, an inviolate territory, entrepreneurs and agents who distribute the national product" (Touraine 1997, 7).

Some Chinese citizens who participated in Tiananmen Square as well as other dissidents agree with Touraine. Recently, they posted two declarations in English on the Internet (Eckholm 1998), addressing civil rights, freedom, and social justice. Writers of the declaration believe its appeal will be broad in China, where the economic boom has begun to taper off. In one of the resolutions they relate justice to markets. "Society's wealth is not being accumulated by hard-working, pioneering and law abiding people, but it is falling into the hands of the dregs of society, the holders of privileged power and their sychophants, corrupt officials and those who steal what they are entrusted with protecting" (Eckholm 1998, A3).

The only reason pundits could confuse the concepts of Chinese open markets and democracy is because the two are hopelessly conflated in postmodern discourses. Furthermore, even though the discursive space of the idea of democracy has been attenuated and is sparse, the sign of democracy itself has an unprecedented broad reach; its signifiers are found in surprisingly disparate discourses. An open market or attempts at an open market does not a democracy make. Indeed, in China and elsewhere (e.g., Burma), talk of democracy in the face of political disenfranchisement of many citizens is evidence of the attenuated nature of this sign in the postmodern world. In some places democracy's discursive space has been ratcheted down to few concepts, but like a thin oil, it spreads across many national regimes.

Technology Discourses

Age of Machines

It is almost impossible for our contemporary freedom fighters to use the rhetoric of a democratic discourse without incorporating some technological concepts, because those concepts were constructed within the articulation of that discourse. Modern technology discourses, surprisingly, have weathered the modern-postmodern transformation almost unaltered and certainly undiminished; they have, in fact been magnified and expanded. Mechanistic processes and principles that form technology discourses remain the same, even as machines and systems theories become more sophisticated. Before the end of the nineteenth century the term "technology" did not exist in its modern incarnation, but referred generally to the practical arts. Without ever using the word "technology," Carlyle at the beginning of the nineteenth century formulated what I believe to be the categories of a modern technology discourse. In a seminal essay, "Signs of the Times" (Carlyle [1829] 1957), he describes machines with both material and mental referents and constructs the sign, "machine" or "mechanism," to signify a specific group of concepts. They are a mechanical philosophy associated with Descartes, Locke, and Newtonian physics; the industrial arts; the systematic division of labor; and an impersonal, hierarchical, bureaucratic organization (L. Marx 1994). Carlyle christened the nineteenth century "the Age of Machines" and indicted the practice of law, the pursuits of science, the institutionalizing of art, and even the development of calculus as all walking the narrow path of systematic procedure and rigid logical thought (Carlyle [1829] 1957). One of the first writers, I believe, to expand the material sign, machine, to include the social and cultural formations that grew up around it, Carlyle ranted about this "Age of Machinery." He left readers with what has become in popular and scholarly technology discourses, a dystopian diatribe, and "Signs of the Times" is remembered for that contribution. It would be as foolhardy to dismiss his diatribe as it would be to embrace the Tofflers' (1995) utopian, but acultural, vision of the Internet.

Carlyle's style is satiric as was Swift's, but his prose is more flamboyant. He writes in the dwindling years of the British romantic literary period. Whereas early romantics such as Blake and Byron created poetic diatribes that were aimed at specific machines and manufacturing processes, Carlyle, as a historian, addressed machines and the discursive space that circulated around machines. In his view, machinery was a master trope enveloping those social, cultural, and intellectual orbiting concepts. Using "systems" metonymically, Carlyle chose it to signify, as it does today, a part of a technological discourse. When the trope was employed in realms to which it was originally alien, it always brought with it intimations of bureaucracy, instrumental reasoning, and a mechanical philosophy.

Warning us about using metaphors such as "social system" or the "machine society," Carlyle says, "Considered merely a metaphor, all this is well enough; but here, as in so many other cases ... the shadows we have wantonly evoked stand terrible before us and will not depart at our bidding" (Carlyle [1929]1957, 28-29).

In "Signs of the Times," Carlyle carefully selects concepts he sees cohering around "machinery." He has the luxury of distance in time from the expectations of the early Enlightenment writers; he sees the effects of machinery, notes the discrepancy between the expectations and the delivery, but still sees the combination as materially and discursively powerful in the articulation of new democracies.

Carlyle's influence. "Signs of the Times" is prescient, however, and contains not only the structure, but foreshadows the claims of most technology criticism to come. Carlyle attacks Bentham and excoriates "the great art of adapting means to ends" (Carlyle [1829] 1957, 22). In a prophetic statement he says, "Everything has its cunningly devised implement, its pre-established apparatus; ... Thus we have machines for education: Lancastrian machines; Hamiltonian machines; monitors ..."(23). Locke he says, is neglecting metaphysics, and "His whole doctrine is mechanical, in its aim and origin, in its method and result"(Carlyle 1957, 23). Horkheimer and Adorno (1972), Marcuse (1968), Ellul, (1964), Postman (1992), and, to some extent, Weber ([1905]1996) inherit a literary form from Carlyle, an essay that is a diatribe against a nexus of technological concepts. But I would like to make an additional point: Both dystopian and utopian critics were granted a legacy of a deterministic discourse containing a set of material and ideational categories, around which their discussions would rotate in the nineteenth and twentieth centuries. Unfortunately, such deterministic texts are indicative of most old and new technology discourses which are themselves usually diatribes or paeans.[3] Although Carlyle created this nexus of technology, he was bequeathed the concepts separately from eighteenth-century scholars who identify progress as history and discursive subjects as romantic, powerless captives in the face of machines. Likewise, he adopts their belief in determinism, which now permeates most technology discourses.

Carlyle's predecessors. Carlyle's nexus of concepts was not formulated alone. He had the advantage of reading early Enlightenment writers introducing the exact topics he selects to identify as characteristic of his century. He most likely read Turgot's 1750 "Discourse on the Progress of the Human Spirit," which many cite as the essay opening the Enlightenment and de Condorcet's "Sketch of a Historical Drama of the Progress of the Human Spirit," which many think signals its close (Williams 1994); he probably read most scholarship in between. These influential essays are themselves based on technological determinism; Turgot and de Condorcet tie the theory of inevitable human progress, and consequently modern definitions of history, to technology (Williams 1994). Turgot writes that historical progress is determined by the creation of systems of transportation and communication across space. "The inconclusive ebb and flow of history would have continued indefinitely had it not been for the crucial inventions that finally overcame historical entropy and pushed the human race onto the track of cumulative progress" (Williams 1994, 223). Although Carlyle will stand in opposition to -Turgot's optimistic tone, he mirrors Turgot's epistemology.

Applied Science Is Progress and Progress Is History

> There is no end to machinery. ... We remove mountains, and make seas our smooth highway; nothing can resist us. We war with rude Nature; and, by restless engines, come off always victorious, and loaded with spoils. (Carlyle [1829]1957, 22-23)

Carlyle's presumed knowledge of Enlightenment writers allows one to understand how his melange of concepts emerged, but the connection is more important than just the Carlyle tie. Noting the appearance of technological ideas in these early texts allows us to see the unfolding of a modern technology discourse within the articulation of democratic ideals.

Before the Enlightenment, historians offered teleological chronologies as their narratives of human life, but the introduction of the modern belief that humans could master unruly nature forever changed their epistemology. Efforts to tame nature were associated with the improvement of the human race and that improvement or progress became the modus operandi of recording history. Turgot is one of the first writers to express this theme when he notes that history is a record of progress driven by the application of science-based knowledge (Williams, 1994), and not only Carlyle, but other influential scholars of the nineteenth century kept that belief intact.

Even Karl Marx fell in step. "In their [Marx and Engels'] view, the critical factor in human development—the counterpart in human history of Darwinian natural selection in natural history—is the more or less continuous growth of humanity's productive capacity" (L. Marx 1994, 250).

Technology Is Revolution

It is obvious that it takes technology to apply "science-based knowledge" or to expand "humanity's productive capacity," but less apparent is the relationship of technology, revolution, and progress.

> For Turgot, intellectual, technical and political revolution are virtually synonymous. ... [He assumed that] the construction of technological systems of communication and transportation to disseminate scientific learning is a political act, for these systems are the weapons that will make the triumph of the bourgeoisie inevitable. Technology is revolution. (Williams 1994, 224)

De Condorcet and later Marx reasserted Turgot's belief in technology as a revolutionary force. Marx understood that the proletariat would have to seize state power and the means of production by revolution. "Technologies of communication and transportation will conquer aristocracy and history itself" (Williams 1994, 225).

Marx's grasp of the power of communication and transportation technologies is foresighted, and, among the nineteenth-century scholars whom I have studied, his approach is the most nuanced. In many nineteenth- and even twentieth-century scholarly discussions of technology, discursive subjects are passive in the face of invincible machines. Even though Marx comprehends the manner in which technologies can imprison or free workers, he assumes that technology is the power that will lead to communism (Cohen 1978). His popular premise about the manner in which machines alienate workers actually appears in his description of the latter stages of an industrial revolution. When addressing earlier stages Marx notes that automation was not the first cause of the development of capitalism, nor the alienation of workers, since at that stage machines were still a means of transmitting human action. His concept of automation, Bimber (1994) observes, was not devoid of the influence of social history. I would say that Marx still allowed space for his ideal discursive subject to act, to have agency. That Hegelian subject was transformed by Marx, who envisioned humans as driven to satisfy basic needs through material means and to resist alienation from their own labor.

> A set of human attributes provides the logic of development of the forces of production and their primacy over other features of social and political life. Another set of environmental conditions facilitate development at various points in history. Technology is among them. (Bimber 1994, 97)

During the second stage of industrialization, however, Marx does not provide his subject with agency. With the division of labor and the further mechanization of work, his subject becomes alienated. This occurs in part because the means of production have been transformed into fixed capital in the form of machines, and that capital accumulates (Bimber 1994). In this stage the workers are powerless in the face of machines.

Now Technology Is Progress

Historian Leo Marx echoes Karl Marx's gloom about the latter stages of an industrial revolution. Speaking not about the human condition, but of the effect of combining the concepts of technology and progress, he states

> The growing scope and integration of the new systems made it increasingly difficult to distinguish between the material (artifactual or technical) and the other organizational (managerial or financial) components of "technology." At this time, accordingly, the simple republican formula for generating progress by directing improved technical means to societal ends was imperceptibly transformed into quite a different technocratic commitment to improving "technology" as the basis and the measure of —as all but constituting—the progress of society. (Marx 1994, 251)

If Enlightenment writers[4] saw technology as that which would improve the human condition by freeing workers from drudgery, and if this liberation was considered progress, than the shift mentioned in this quote is foreboding. The nineteenth-century Marx (Bimber 1996) and the twentieth-century Marx (1994) note the shift of the denotation of "progress" from the improvement of human life (even if by applied science) to the accumulation and improvement of machines. This perception remains today. Nineteenth- and twentieth-century scholars (Carlyle, Marx, Veblen, Horkheimer, Adorno, Marcuse, Ellul) offer various explanations for this transformation, but most explain the dominance of a technological discourse by attacking the evils of capitalism, for example, open markets, the establishment and greed of the managerial class, and the division of labor.

Carlyle's 1829 technological discourse remains intact during this time, but aspects of it are sharpened by Weber at the close of the nineteenth century when both Weber and Veblen introduce readers to the term "technology." This new term incorporates Carlyle's material and ideational referents. Weber,[5] opposing Marx and the formation of socialist governments, attacks capitalism as Marx had, but it is a different assault. Whereas Marx constructs a material but ideal subject driven by the need to accumulate and to resist alienation from self, Weber constructs a reactionary ideal subject, one who valued premodern Christian norms and tradition. In Weber's text ([1905] 1996) these ideal humans see divine validation in personal wealth and allow a theocracy to establish norms and social practices. Whereas Marx thought the long-term effect of capitalism would be to alienate humans from their true selves, Weber believed the long-term effect of capitalism would be to alienate humans from their culture and traditions. In this view, Christians of the late nineteenth and early twentieth centuries who believed in instrumental rationality instead of following tradition were false Christians, while in Marx's view, those who could not see that the investment in machines and capital would alienate humans from themselves had false consciousness.

Weber's major contribution to the ongoing technological discourse was to clearly illustrate the role of instrumental rationality. In his view it dominated most concurrent social and political discourses and positioned passive subjects in those discourses. Building on the idea of Carlyle's belief that the philosophies of Descartes, Locke, and Newton were mechanistic at best, Weber ([1905]1996) declared that an instrumental mode of thought which confuses ends and means was a product of capitalism, specifically bureaucratic practices within capitalism. He states that society had "complete dependence on its whole existence, of the political, technical and economic conditions of life, on a specially trained organization of officials" (Weber [1905] 1996, 216). Weber's position was adopted by twentieth-century scholars (Heidegger 1977; Horkheimer and Adorno 1972; Marcuse 1968; Habermas 1968; Ellul 1964; Schmitt in McCormick 1997; and Mumford 1934) who, by and large, keep Carlyle's nexus of technological concepts intact. These critics consistently warn us about the dehumanizing effects of political systems that are governed by technological designs and instrumental thought.

Technology Discourses with Democracy

Since technological discourses unfolded in democratization efforts, and since a modern technological discourse remains intact today, one might expect traditional technological concepts to be present in the new discursive space of the resistance stories in this text. Certainly their belief in the ability of machines to advance their causes may indicate underlying assumptions of instrumental rationality, but other hallmarks of technological discourses are seldom present in their rhetoric. Discussions of standardization, systematization, and impersonal hierarchical organization are absent from the accounts here or the initial narratives of other resistance movements. And even though rhetorical logic is not absent from their stories, their discourses are not ruled by instrumental rationality. They are influenced, it seems to me, by eighteenth- and early nineteenth-century democratic rhetoric. Like these early texts, their discourses disclose an ideal citizen, a subject who has certain values, ethics, knowledges, privileges, and so forth. Their subjects have different ethnic and racial identities, but some values, attitudes, and understandings of these subjects are similar, no matter their national identity. One might say, therefore, that a discursive site of major social and political power—a technology discourse—was partially bypassed in the establishment of new sites of power. It is important not to confuse a technology discourse in these cases with the information technology itself. These resisters appropriated electronic communication technologies and garnered power by capturing and controlling some of the flow of information. What was seldom seen in their rhetoric was Carlyle's nexus of concepts, which still permeate sites of corporate and governmental power today.

The ascendance of democratic, rather than technocratic, goals may be a function of the optimistic nature of grassroots groups entering the political arena. As their size increases and they gain fiscal maturity, turning their attention to economic as well as social forms of government, this ascendance may not remain in their stories. It is, nonetheless, a phenomenon of their early tales. One might call this new discursive space an amalgam of the accounts of contemporary social conditions and historical democratic concepts. It would be helpful, however, to ask more specifically just who these resisters are.

Who Are They? Subjectivity and Identity

These freedom fighters are seeking, as I have said, modern democratic goals, but operating in a postmodern, post-Berlin Wall environment. Moving between modern and postmodern techniques, I shall try to describe the freedom fighters and coalition participants included in this text and in other accounts of resistance movements around the world. To relate the narratives of authors here to stories of other revolutionaries, I draw upon two scholars, Alain Touraine and Manuel Castells, who write extensively about resistance movements. Even though the concept of subjectivity occupies center stage in postmodern discourse analysis, it is highly contested within various forms of this analysis. And since the concept has such explanatory power, it has been appropriated by diverse modern scholars who may wish to demote the use of objectivity in their analyses. In this arena subjectivity is also a controversial signifier. With this situation in mind, I will start to answer my question, "Who are they?" by addressing subjectivity.

Subject

A subject, as I understand it, is a paradoxical concept, intended to be double-edged. It emerges in discourse as speakers or authors communicate their ideas, desires, values, assumptions, knowledge, and so on. Speakers or authors can be the subjects of their meanings. In other words, they create and control their own utterances, but those meanings have been socially established. Specific interpretations belong to larger social discourses. Speakers are members of communities which incorporate these discourses; therefore, they are subjected to the values and assumptions of discourses. In fact, the rhetoric of a community to which speakers belong may be so naturalized as to be transparent, and often they do not recognize the constructed nature of an utterance. The discourse, therefore, speaks through them. They are subjected to or enslaved by the discourse; that subjects can be rulers and slaves of meaning in discursive space is the paradoxical nature of subjectivity.

A shortcut to ascertaining subjectivity in a communication was established by Elizabeth Ellsworth (1990, 1997) when she posed the question, Just who does that text think its readers are? Forms of address and other cues within the text can reveal the manner in which the communication thinks of and constructs a human. Reader theorists Jauss (1982), Iser (1978), and Fish (1980) all agree that the text holds clues to ascertain the subject.

In this chapter and elsewhere (DeVaney 1998), I have argued against the romantically ideal or transcendentally ideal subject who appears in pro- and anti-technology discourses, but the construction of any subject in a text fixes that idea of a person. The subject of a discourse once posited has specific knowledges, believes in certain ethics, and embraces specific values. And while some postmodern inquiry has expanded the boundaries of subjectivity, a subject by the very imprisoning nature of language has parameters and is established. In other words, subjects can only possess the characteristics of some ontological notion of a human. Subjects might be ideal, romantic, technocratic, totalitarian, fascist, and so on, but placing a subject on the page of a text determines her characteristics. My point here is that the critique of an ideal subject in a technological discourse should not be always construed as negative. One would hope that a discourse could make space for multiple subjects and subjects who respect diversity, whether it be ethnic, racial, or religious; and some contemporary discourses do open up the space for subjectivity by welcoming multiple subjects. Some critics (Butler 1993) have located subjectivity in discourses in which the sex role, for instance, is a performance; that discursive move allows the concept of sex to be less determined, yet not truly fluid.[6] The subjects of discourses in this text and in resistance narratives from around the world have specific boundaries and parameters that I shall discuss later.

The two authors who have guided me through the issues of democratic resistance movements have differing comprehensions of subjectivity, and in turn both of their understandings differ from mine. Since the scholarship on resistance movements is so new and as yet underdeveloped, all approaches to subjectivity, modern or postmodern, shed light on the implications for these movements. Few scholars are writing about these revolutionary events and fewer still have developed a consistent line of inquiry into the meaning of resistance movements. Of the very few who have, I found that Castells and Touraine appear to be the most thoughtful, and to have produced the scholarship most salient to the issues considered in this book.

Psychology and Sociology

The manner in which Touraine and Castells approach subjectivity and identity is complex and informed by the history of theories in several academic disciplines. The concept of "identity" has a longer history than subjectivity, and this history adds to the confusion of its use by Touraine and Castells as a sociological concept. Identity, originally tied to the idea of self, was introduced to psychology by William James, but more fully articulated in sociology by Cooley, who conceived of the looking-glass self (Baumeister 1987). He argued that the self was a reflection of how people saw us. The work of Cooley and later George Herbert Mead was the basis for the theory of symbolic interactionism, which essentially defined the field of sociological social psychology (Baumeister 1987). Symbolic interactionism,which addressed identity among other things, isn't as important here as is the fact that two academic disciplines merged interests and concerns with the establishment of sociological social psychology. The distinction between the psychological study of self or personal identity was blurred with the sociological study of group identity. This blurring is one of the issues I grapple with when reading Touraine and Castells.

Touraine on Subjectivity

In his description of the subject, Touraine employs the rhetoric one might use to delineate personal identity. He alleges that the construction of the subject is the desire of an individual to create a personal history and give meaning to the experiences of her or his own life. Although the lines are blurred here between the realm of the personal and that of the social, Touraine does note that the effort to give meaning to personal experience can only be realized socially. In the same paragraph, he states that the subject is established, "not alone, but in the company of those groups fighting against autocratic regimes where totalitarian subjects reside within the hegemonic economic centers, and within international financial systems" (such as the global market) (Touraine, 1995, 28). His subject as a member of a group "must do double combat against the incapacity of most communities to stop the valorization of markets and against the world of commodities itself" (Touraine 1995, 29).

Touraine attempts here and elsewhere (1990, 1997)[7] to explore the paradoxical nature of the subject that frees and imprisons, by noting the tension between the highly regulated and unregulated spaces in which people move. He elaborates his description of subjectivity by perceiving its establishment in solidarity, that is, groups fighting against commodification and the market. But here and in *What Is Democracy?* (1997) he envisions a space outside of language for the formation of thought to construct ideas of self and personal identity and then relates that space to subjectivity. "To the extent that the subject is its own creation, the social actor is self centered, rather than socially centered; the subject is defined by its freedom, not by the roles it plays" (Touraine 1997, 124). For him and some other sociologists there exists unmediated experience that feeds the conception of self.

In any case, all individuals are more continuously caught up in relations of dependency or cooperation than in linguistic exchanges. An individual works, commands, or obeys and encounters scarcity or abundance. ... Knowledge of the other is always preceded by the quest for the self. ... (125)

Castells on Subjectivity

Likewise, Castells (1997) searches for that space outside of discourse when describing identity and the meaning of identity.[8] He asserts that shared experience is the sine qua non of group identity. Whereas shared experience may be the key to group identity, it is only understood, I believe, through the mediation of language. The students and citizens marching in Tiananmen Square understood the meaning of their actions and experience because they had communicated to one another desires for freedoms of speech and the press. In fact, they used selections from the writings of Abraham Lincoln to crystallize their intent. Speaking among themselves[9] before, during, and after the events, they gave meaning to a new discourse and their rhetoric positioned them as subjects in that discursive space.

Castells does not ignore the concept of subjectivity in favor of identity, but he does demote it to a subcategory of one of his types of identity. He establishes three descriptions of identity for citizens within a democratic state, namely legitimizing identity which can be seen, he says, as a mode of state domination in the eyes of some, or a means for civility in the eyes of others; identity for resistance in which communes or communities are formed; and project identity in which citizens build new identities that redefine their position in society. About his legitimizing identity he says,

> The conquest of the state by the forces of change (let's say the forces of socialism, in Gramsci's ideology) present in the civil society, is made possible exactly because of the continuity between civil society's institutions and the power apparatuses of the state, organized around a similar identity (citizenship, democracy, the politicization of social change, the confinement of power to the state and its ramifications and the like). Where Gramsci and de

Tocqueville see democracy and civility, Foucault or Sennett, and before them Horkheimer or Marcuse, see internalized domination and legitimation of an over imposed undifferentiated normalizing identity. (Castells 1997, 9)

I shall return to his second type of identity later, but it is within his third process of constructing a project identity that subjects are formed. In fact, he says "subjects, as defined by Alain Touraine" (Castells 1997, 9). But later, Castells presents a caveat, "Subjects are not individuals, even if they are made by and in individuals. They are the collective social actor through which individuals reach holistic meaning in their experience" (10). It appears that he wishes to stress group identity even though he admires the way in which Touraine (1995) highlights the desire of an individual to create a personal history and give meaning to the experiences of her or his life. The psychological and social approaches of Castells's and Touraine's ideas of subjectivity affect their visions of resisters.

Touraine's claim that subjectivity is the desire of an individual to create a personal history resembles the psychological idea of self-identity, even though he labels it subjectivity. And Castells's claim that he borrows Touraine's notion of subjectivity, but that "subjects are not individuals. ...They are the collective social actor" (Castells 1997, 10) resembles the sociological idea of group identity. I say this because their explanation that subjectivity is formed outside of discourse and their description of subjectivity more closely resemble the modern psychological and sociological notions of individual and group identity.[10]

Subjects in Resistance Movements

In an attempt to understand the agents of contemporary resistance movements as subjects, Touraine states, "Democracy is the battle waged by subjects, in the context of their culture and their liberty, against the domineering logic of systems" (1997, 12). The discourses in this text, I believe, seldom articulate the nexus of concepts forming a technological discourses, and although I am certainly in Touraine's debt, I disagree with his description of these subjects. By invoking the "domineering logic of systems," Touraine uses the phrase metonymically to stand for a technological discourse that includes the range of concepts I identified in Carlyle. In this mode, Touraine continues what has now become a time-honored tradition of pitting a technological discourse against battles for democracy. In other words, democratic subjects (or those fighting to achieve that status) must divest themselves of, must be free of technological discourses. I do not think, as does Ellul (1964), that the fight for certain freedoms is simply one against systemization; it is insufficient to explain the shackling of citizens. Just as the "logic of systems" is used by Touraine and Ellul as a metonymic trope for a technological discourse, so too do they conversely suggest that a technological discourse is the trope for those undemocratic practices currently shackling citizens. That claim again attributes too much power to a technological discourse, making it a causative agent. In appropriating the dystopic rhetoric of certain traditional discourses, contemporary scholars are also subjected to romantic claims. And while it is tempting to blame a technological discourse, which might actually be one of the mechanisms used to achieve nondemocratic ends, it is insufficient; no one discourse would have that power.

Agency

One might say that Touraine would not be surprised to find the diminution or absence of a technological discourse in the chapters of this text. He identifies those resisters who sue for circumscribed rights as active agents who have rejected major revolution and the modernization it demands. If I reject that explanation, what might explain the lack of a technology discourse in the democratic resistance stories whose subjects appropriate technology? Is it an epiphenomenon that it was bypassed? Differential treatment of subjects in revolution and resistance movements might provide clues. Touraine might claim, as do many scholars, that citizens are blind to the logic of systems, but it runs counter to his assertion about the fact that many contemporary resistance fighters have agency, to him they are active, not passive, subjects. He makes a distinction, as I noted earlier, between those early revolutionaries who sought to overthrow governments and establish democratic states, and some contemporary revolutionaries who work for smaller changes such as additional egalitarian social practices within established governments.

The former, he argues, sought radical social change for purposes of modernization which eventually enriched classes of people, not individuals. He believes these early revolutionaries were subjects of and, consequently subjected to, technological discourses that mandated modernization, and pictures them as passive actors in the thrall of an instrumental discourse. Touraine also fears that a revolution that seeks to overthrow a government, instead of just change some practices under that government, ultimately opens itself up to the same injustices practiced by the former government. While Castells agrees with the inability of states to handle religious or nationalistic movements, he uncovers strength in what I call the "attenuated democratic space." But resisters who now fight only for specific freedoms are, to Touraine's way of thinking, agents or active subjects, because they seek neither to overthrow nor to modernize their government. This work, he believes, guarantees their agency.

In a post hoc analysis of early revolutions, it is easy to understand how they were informed by Enlightenment ideals that included technological goals and discourses; in fact, I have spent quite some time talking about that here. It is also easy to see how managerial and other bourgeois classes were formed in new republics, but I find Touraine's distinction between old passive actors and new agents troubling. It is as difficult for me to accept the description of passivity in his argument as it is to accept it in dystopian or utopian technological debates. While my vision of humanity does not exclude passivity in the face of overwhelming social forces, my vision of revolutionary participants and writers, old or new, is just the opposite.

Race

Early democratic discourses originally crafted to overcome oppression fill some of the new discursive space created by the resistance movements. The irony, however, of some people of color appropriating classical democratic rhetoric is not to be missed. I agree with Touraine when he says that, "Democracy's greatest task is to produce and defend diversity within a culture" (Touraine 1997, 12), but in those early democratic texts, the concept of diversity is absent. The equity so prized and sought by eighteenth-century authors was equity for white men only. In their works there is a unified subject, the white man. Western writers constructed a discursive subject by universalizing their conception of a human being and creating a transparent globally unified subject, the Western white male. Discourse analysis often allows that which is rhetorically transparent to become opaque. And from what I remind myself is a position of privilege, I and others can deconstruct the early texts.[11] It then becomes obvious that the modern description of the democratic subject did not account for culture, race, gender, and ethnicity.

But classical democratic texts that guaranteed equality for all and delivered equality for white men (Wiebe 1995) are being appropriated, nonetheless, by nonwhites. Such a move opens up discursive space and allows room for multiple subjects in the new resistance discourses. On a smaller scale this occurred during the civil rights movement in the United States.

Recent Chinese dissidents have opened the issue of the universality of democratic rights for any person in any country. Jiang Peikun, the chief drafter of the September declarations I cite above, was the father of one of the Tiananmen Square students killed in the uprising. Speaking for all the drafters of the Internet resolutions, he said that individual freedoms and rights were a core value of all humanity (Eckholm 1998). While intellectual battles rage in the academy about the inequitable nature of totalizing statements, even here some scholars (Rorty 1998; Touraine 1995; Benhabib 1992) have returned to the modern belief that certain liberal ideals such as human and civil rights are universal.

> "They are who they say they are."
> (Castells 1997, 70)

I'd like to return to Castells' three types of identity, because I believe the freedom fighters of this text and elsewhere would fall into his second category.

> The second type of identity building, identity for resistance, leads to the formation of communes or communities in Etzioni's formulation. This may be the most important type of identity building in our society. It constructs forms of collective resistance against otherwise unbearable oppression, usually on the basis of identi-

ties that were apparently clearly defined by history, geography, or biology, making it easier to essentialize the boundaries of resistance. (Castells 1997, 9)

Not only would the freedom fighters fall into this category, but as I indicated before, their subjectivities are partially fixed by what Castells calls their histories and geographies. These similarities create aspects of their physical boundaries, but I believe that the plea for global help with their causes, by means of contemporary information technologies, has opened discursive space for citizens with other histories and geographies to participate.

Concerned citizens outside the geographic sites of resistance communicate and promulgate the fights for liberty, grabbing the attention of news readers, media watchers, and Internet users around the world. They organize coalitions and offer financial aid. The new discourses, therefore, of freedom fighters and coalition participants have space for multiple subjects by virtue of the fact that information technologies have been used to communicate their plight. Speakers in different physical spaces have been invited to join the cause and extend its reach. In the American Revolution, the French offered moral and financial support only after a lengthy visit by Benjamin Franklin. American revolutionaries were officially supported by the French state itself, but current global subjects in the new discursive spaces identified here are mainly self-selected and can only seldom muster corporate and federal support for their foreign crises.

Space for multiple subjects. It would be foolhardy to claim that cyberspace has erased physical boundaries, making us all potential subjects, let's say, in the Burmese fight for political freedoms. Castells wisely reminds us that similar experience creates similar group identities. Citizens in countries foreign to the resistance, those who do not share the geography or history of resisters, occupy different space in the discourse, but share similar values. They are adjuncts, assistants, aides in the fight; my claim is that cyberspace has allowed multiple subjectivities to emerge in these discourses.

Civil Rights Movements

In addition to his general discussion of identity, Castells (1998) speaks specifically of contemporary independence movements by identifying them as heirs to the wisdom of 1960s and 1970s social and cultural movements. Civil rights fighters in the United States or the students in Paris in 1968

> intuitively knew that access to the institutions of state co-opts the movement, while construction of a new revolutionary state perverts the movement. ... They were essentially cultural movements wanting to change life rather than seize power. ...[Like today's freedom fighters] their ambitions encompassed a multidimensional reaction to arbitrary authority, a revolt against injustice, and a search for personal experimentation. (Castells 1998, 339)

Certainly this description of the goals of some current resistance movements (excluding South Africa and Russia, who replaced governments) is insightful and would grant those resisters, as does Touraine, an intelligence and agency borne of their ability to understand history. Contemporary freedom fighters may have modeled themselves after the civil rights fighters of the 1960s and 1970s. Certainly, they were emboldened by the precedents and partial success of these early movements, and they probably believed, as did United States student resisters in their 1968 demonstration at the Democratic National Convention in Chicago, that "the world is watching."

The dominant discourse of these contemporary freedom fighters, I have said, is democratic and bypasses most aspects of a technological discourse, even though they have applied technology. Perhaps grassroots movements[12] outside of governmental, military, and financial sites of power do not engage the central concepts of a technological discourses, but there may be another explanation.

Castells notes a discursive phenomenon similar to that which I have observed here, "And while they [1960s and 1970s freedom fighters] coexisted, broadly speaking, with the information technology revolution, technology was largely absent from the values or critiques of most movements, if we except some calls against de-humanizing machinism and their opposition to nuclear power"

(Castells 1998, 340). Discursively, Castells is indicating that the accounts of the 1960s and 1970s movements, like the narratives I cite, bypassed aspects of a modern technological discourse. If that is so, the current independence movements are certainly the heirs of the older social and cultural resisters.

New Subjects, New Definitions

The scenarios are complicated. In this text and elsewhere, both strange and familiar modern and postmodern democratic movements are creating new physical and discursive space. Resisters are appropriating new digital communication technologies usually controlled by corporate or governmental agencies in the traditional space for technology. Decentralized technologies are in the hands of the heretofore powerless who have only had roles as consumers in the modern technological discursive space.

Democracy or? The fact that the rhetoric of grassroots dissidents fighting for democracy resembles that of early democratic scholars is telling. Although contemporary freedom fighters have opened discursive space around the concept of "man" in the phrase "the rights of man," they have not (with the exception of the South Africans) included what some United States social scholars accept as diversity. The discursive expansion of "man" in their communication has occurred of necessity because of the race or ethnicity of the participants in any one nongovernmental organization (NGO). In other words, it is still not clear that in Malawi or Indonesia, for instance, the rights sought for dissidents will be shared with all citizens, regardless of race, ethnicity, or religion. Certainly in NGOs that coalesce around nationalistic agendas, "rights" are sought for their constituents only. But it is unclear whether the attenuated democratic rhetoric of some resisters exists for purposes of publicity and for establishing ties with wealthy Western democratic citizens, or if a fully articulated democracy is the actual goal of these resisters. Under the gaze of the state, their articulations must be guarded and calculated. To be heard in the West, they must use language that expresses values embodied in Western beliefs (and it is this decision that may further narrow their subjectivity within their discourses). Yet who can argue with the granting of rights to oppressed citizens? So the utility of this discursive move in public is that utterances may in an ethical arena take precedence over the circumscribed subjectivity of their democratic discourses. It will be vital to watch the trajectory of the movements, especially those who experience success, to see what is the nature of subjectivity in their continued discourses.

Equality or? The other aspect of the rhetoric of global dissidents that remains unified in each of their discourses is the concept of equality. The discursive space of that concept in early democratic United States texts included enfranchisement, educational and job opportunities, access to the ownership of private property, and the guarantee of certain rights such as freedom of speech, assembly, and the right to bear arms, among others. I extract the rhetorical promises from those early texts and explore the manner in which the concept of equality was conceived to assess the manner in which contemporary dissidents conceive of the concept of equality.

In both scholarly and public discourses the original use of the term "equality" has been appropriated in divergent democratic agendas and today is a hydra with many heads. Sen (1992) details the search for a class of equal entitlements sought by libertarians, or the demand by utilitarians for equal weight on every unit of utility, or the requirement of equal incomes from the "income-egalitarians, or the call for equal welfare from the "welfare egalitarians." His analysis uncovers the actual social inequality created in the wake of fights for certain equalities constructed within traditional social and political agendas.

What appeared to be a full explanation of equality in early democratic texts has over the years been exposed as insufficient. Yet the global dissidents of whom I write seek a still more limited form of equality than that presented in the United States Constitution, for example. Again, this discursive strategy may in the long run prove to be the most expedient method of promulgating their messages. The aim is, after all, to make their voices heard around the world and to attract those democratic citizens whose beliefs in equality resides in the way that concept is articulated in contemporary political discourses in the United States, for instance. Similar tactics, in which equality was narrowed to address specific rights, were used in the civil rights movements in the 1960s and are certainly effective ways of motivating citizens to vote or join alliances with oppressed peoples. Yet in constructing a constitution, such as that of South Africa's Government of National Unity, and writing legislation, the complexity of the concept ultimately needs be addressed.

Political expediency is evident in the rhetoric of the grassroots movements of which I write. Their subjectivity is circumscribed and their freedoms are few and well defined. Both Castells and Touraine have said that the success of groups combating their governments in a battle for specific rights is far more likely to produce a democratic environment than is the complete overthrow of a government. If they are correct, the expedient and ethical discursive strategies of the contemporary freedom fighters have chances of success despite their existence under the gaze of the state.

The freedom fighters and coalition participants described in this text were not co-opted by technology discourses, as critics say we will all be; systems logic, standardization, and impersonal hierarchies are largely absent from their accounts. They did not express deterministic beliefs; they were not passive in the face of large electronic networks or ruling regimes. They were agents acting to alter, transform, or resist social and political practices around them.

Their narratives constructed an ideal subject imbued with values, ethics, and knowledge from early democratic texts.[13] Moving beyond the essential boundaries of those early texts, some of them opened up discursive space for diversity, for multiple subjects. With the assistance of communication technologies they created new subjectivities for global citizens who shared their values, but not their histories nor physical geographies. They expanded their discursive space, their virtual geography, online. Many may have learned from the civil rights movements in the 1960s and 1970s and sought, consequently, to gain limited freedoms, to fight in an attenuated democratic space. Others established new governments with fully articulated democratic goals.

With their global allies these resisters gained partial control over the flow of information in their countries. Such a move yoked technology and resistance in a manner that will hopefully persist for decades to come. If these global allies become the watchdogs and gatekeepers of new rights, freedoms, and liberties, their subjectivity expands beyond these initial resistive discourses. Burmese expatriates in the United States and Chinese dissidents in China recognize these global gatekeepers. Nelson Mandela believes they will persist and said as much in a farewell speech before the United Nations:

> I will continue to entertain the hope that there has emerged a cadre of leaders in my own country and region, on my continent and in the world, which will not allow that any should be denied their freedom as we were; that any should be turned into refugees as we were; that any should be condemned to go hungry as we were; that any should be stripped of their human dignity as we were. ...Then would history and the billions throughout the world proclaim that it was right that we dreamed and that we toiled to give life to a workable dream. (Mandela 1998, A3).

Notes

1. In this chapter I use the phrase "postmodern world" instead of "New World Order," because many of the grassroots movements reported here and elsewhere stand against the economic structure of the New World Order.

2. I maintain contact with and receive stories from Asian and African students who received their Ph.D.s under my guidance in the late seventies and eighties. Some of them have participated in democratic resistance movements and employed electronic technology along with their colleagues.

3. Heidegger asserted that scholarship that simply investigates whether the objects of study in a field are good or evil can reduce an important line of inquiry to idle chatter (McHoul 1998).

4. The fusion of separate concepts, technology and progress, into a unified myth had a particular genesis in the United States. Although the myth has its origin in Western industrial revolutions and was explained in the rhetoric of Enlightenment philosophers, its articulation in America depended on the acceptance of doctrines emerging from utilitarianism, empiricism, and social efficiency in the earlier part of this century (Berman 1994; Cahoone 1988; Ross 1991; Wiebe 1995). During the second wave of the industrial revolution in the United States, the country experienced its largest capital expansion. Pragmatic citizens saw technology at the heart of this prosperity, and their belief in machines could not even be dislodged by an

economic depression (Sclove 1995). By the thirties, the myth of technology-as-progress had been incorporated as a stable maxim in public discourses.

5. In "The Protestant Ethic and the Spirit of Capitalism of Capitalism," Weber claims "that the shift by Calvinists and some others from a heavenly reward to an earthly reward helped [expand] the capitalistic spirit and caused substantial economic growth in Protestant countries. Later, the original religious attitudes were transformed and replaced by the secularist capitalist ethic. By extension, the spirit of capitalism spreads out from the early centers and has sweeping general effects" (Goldstein and Boyer 1986, 438).

6. The tension between identity and subjectivity is uncovered by Derrida and deMan when they critique the politics of identity as it appeared in 1980s cultural studies. Within scholarly analysis of some cultural issues they find problematic the positing of a subject fixed in a biological construct of race or gender. Derrida and deMan thought the stability of the sign, such as Chicana, pushed readers toward essentialized meaning, whereas they were encouraging everyone to consider the dispersion of language. Butler (1993) went a long way toward improving postmodern scholarship conducted in the name of identity politics when she noted the constructed nature of both traditional signifiers of woman, namely "gender" and "sex." Since de Beauvoir (1953), feminists had agreed about the social construction and fluidity of gender; but not until Butler, in a Derridean move, identified the socially produced nature of sex, did readers consider it anything but fixed. Describing sex as a linguistic concept that must be formed and only formed discursively, Butler set about describing those discursive practices and performances that constructed the concept of sex. With this move she helped mitigate, not eliminate, essentializing and totalizing claims brought by critics against those who practiced identity politics. That Butler used a strategic Derridean argument to answer his criticisms is deliciously ironic. Derrida (*Of Grammatology*) convincingly demonstrates that the relationship between de Saussure's signifier and signified is not fixed, thereby collapsing the space between the signifier and the signified. In a parallel, not exact, move Butler collapses the space between the signifier, "sex" and what had been considered stable, the signified "sex," and leaves us with a constructed and more fluid concept. I do not ignore here that the source of the critique of identity politics was two white males. Multicultural and feminist texts were receiving much critical attention in the eighties and decentering the white male cannon. In my view, many white scholars were happy to receive the deconstructive critique and acted on it.

7. The description of identity from Touraine is my translation.

8. I am not equating subjectivity with identity here. I simply wish to note the helpful contributions of Castells and Touraine to an understanding of the resisters. To do that I return to their notions of subjectivity and identity, because each of these concepts sheds light on the issues of this chapter.

9. I have been convinced by de Saussure (1974) that language precedes conceptualization, precedes the existence of independent entities, and makes the world and our thoughts intelligible by differentiating between concepts.

10. The fact that they are modern does not diminish their extensive contribution to the literature on resistance movements, it simply uncovers my own theoretical position. As I mentioned above I believe that language precedes thought, therefore, I have a linguistic frame of reference. Also, my understanding of subjectivity has been influenced by my interpretation of, among others, Jacques Derrida, Hans Jauss, Stanley Fish, Michael Bove, Judith Butler, Barbara Herrnstein Smith, and Diana Fuss.

11. As scholars and critics we have always occupied a position of privilege when we ply our trade, the interpretation of texts. On the other hand, while we must respect authors' or speakers' intentions, their writings and utterances take on a meaning of their own and are, therefore, open to explication.

12. In writing about the relation of grassroots movements and people who reside on the borders of countries, Appaduri (1990) believes that they create imaginaries. The discursive geography of countries around the world, he says, has become to some extent, mediascapes and technoscapes by virtue of the global reach of television, film, and new telecommunication technologies such as fax, phone, email and the Internet. His media and technoscapes proffer a way of life to the poor and powerless that is unattainable; these scapes help citizens create imaginary worlds in which their participation is unrealistic. And these imaginaries, he concludes, influence the activity, social life and communication of people on the border; they are caught in a dream. But many freedom fighters in this text and elsewhere were powerless and poor, were subjected to the power of media and technoscapes, and yet they did not act on unrealistic imaginaries. They were not defeated by dreams, but seized control of the flow of some information. They were not passive in the face of powerful telecommunication technologies, nor their corporate providers.

13. Contemporary grassroots movements have used the term "neo-liberal" to describe those New World Order economic and political structures such as the World Bank, the International Monetary Fund, and the United Nations. NGOs working against globalization note that the appropriation of

the concept "liberal" by powerful economic groups and multinational corporations masks the oppressive and inequitable practices of these dominant organizations. The use of the term "liberal" in this view simply hides their need to accumulate capital and rule. Resistors, therefore, label their usage neo-liberal. I have not used that term in this chapter because of the pejorative connotation it now entails, namely corporate greed and oppression. But, I must note that the resistance narratives in this text and elsewhere do contain liberal values.

References

Appadurai, Arjun. 1990. Disjuncture and difference in the global cultural economy. Special issue of *Theory, Culture & Society* 7: 295-310.

———. 1991. Global ethnoscapes. Notes and queries for a transnational anthropology. In *Recapturing Anthropology,* edited by Richard G. Fox. Sante Fe, NM: School of American Research Press.

Baumeister, Rolf F. 1987. How the self became a problem: A psychological review of historical research. *Journal of Personality and Social Psychology,* 52(1): 163-176.

Benhabib, Seyla. 1992. *Situating the self. Gender, community and postmodernism in contemporary ethics.* New York: Routledge.

Berman, Art. 1994. *Preface to modernism.* Urbana: University of Illinois Press.

Bimber, Bruce. 1994.Three faces of technological determinism. In *Does technology drive history?* edited by Merritt Roe Smith and Leo Marx. Cambridge, MA: MIT Press.

Bove, Paul. 1992. *Mastering discourse: The politics of intellectual culture.* Durham, NC: Duke University Press.

———. 1992. *In the wake of theory.* Hanover, NH: Wesleyan University Press.

Butler, Judith. 1993. *Bodies that matter. On the discursive limits of sex.* New York: Routledge.

Cahoone, Lawrence. 1988. *The dilemma of modernity: Philosphy, culture, and anti-culture.* Albany: SUNY Press.

Carlyle, Thomas. [1829]1957. Signs of the times. In *Thomas Carlyle: Selected works, remembrances, and letters,* edited by Julian Symons. Cambridge, MA: Harvard University Press.

Castells, Manuel. 1998. *End of millennium*. Malden, MA: Blackwell Publishers.
———. 1997. *The power of identity*. Malden, MA: Blackwell Publishers.
Cohen, Gerald A. 1978. *Karl Marx's theory of history*. Princeton, NJ: Princeton University Press.
de Beauvoir, Simone. 1953. *The second sex*. London: Penguin.
de Man, Paul. 1986. *The resistance to theory*. Minneapolis: University of Minnesota Press.
Derrida, Jacques. 1985. The ear of the other. In *Texts and discussions with Jacques Derrida*, edited by Claude Levesque and Christie McDonald. New York: Schocken Books.
———. 1978. *Writing and difference*. Chicago: University of Chicago Press.
———. 1976. *Of grammatology*. Baltimore: Johns Hopkins University Press.
de Saussure, Ferdinand. 1974. *Course in general linguistics*. Translated by Wade Baskin. London: Fontana.
De Vaney, Ann. 1998. Can and need educational technology become a postmodern enterprise? *Theory into Practice*, 37(1): 72-80.
———. 1998. Will educators ever unmask that determiner, technology? *Educational Policy*, 12(5): 568-585.
Eckholm, Erik. (1998, September 30). Chinese dissidents issue a sharp challenge to the government. *New York Times*, A3.
Ellsworth, Elizabeth. 1997. *Teaching positions*. New York: Teachers College Press.
———. 1990. Educational films against critical pedagogy. In *The ideology of images in educational media*, edited by Elizabeth Ellsworth and Mariamne Whatley. New York: Teachers College Press.
Ellul, Jacques. 1964. *The technological society*. New York: Vintage Books.
Ferguson, Robert A. 1994. *The American enlightenment 1750-1820*. Cambridge, MA: Harvard University Press.
Fish, Stanley. 1980. *Is there a text in this class?* Cambridge, MA: Harvard University Press.
Goldstein, Jan, and Boyer, John W., Eds. 1986. *Nineteenth-century Europe: liberalism and its critics*. Chicago: University of Chicago Press.

Government of National Unity. 1996. *Constitution*. South Africa. Available at: http://www.gcis.gov.za.

Habermas, Jurgen. 1968. *Knowledge and human interests*. Boston: Beacon Press.

Heidegger, Martin. 1977. *The question concerning technology and other essays*. New York: Harper and Row.

Herrnstein Smith, Barbara. 1988. *Contingencies of value*. Cambridge, MA: Harvard University Press.

Horkheimer, Max, and Adorno, Theodor. 1972. *Dialectic of enlightenment*. Translated by J. Cumming. New York: Herder and Herder.

Iser, Wolfgang. 1978. *The act of reading*. Baltimore: The Johns Hopkins University Press.

Jauss, Hans. 1982. *Towards an aesthetic of reception*. Translated by T. Bahti. Minneapolis: University of Minnesota Press.

Mandela, Nelson. Mandela words a poignant farewell. *New York Times*. 22 September 1998: A13.

Marcuse, Herbert. 1968. *Negations: essays in critical theory*. Boston: Beacon.

Marx, Leo. 1994. The idea of "technology" and postmodern pessimism. In *Does technology drive history?* edited by Merritt Roe Smith and Leo Marx. Cambridge, MA: MIT Press.

Marx, Leo, and Smith, Michael. 1994. Introduction. In *Does technology drive history?* edited by Merritt Roe Smith and Leo Marx. Cambridge, MA: MIT Press.

McCormick, John. 1997. *Carl Schmitt's critique of liberalism: Against politics as technology*. Cambridge: Cambridge University Press.

McHoul, Albert. 1998. Cybernetymology and ~ethics. *Postmodern Culture* 9(1), (http://www.press.jhu.edu/journals/postmodern_culture/v009/9.1mchoul.html).

Mumford, Lewis. 1934. *Technics and civilization*. New York: Harcourt, Brace & World.

Porteus, Kim, and Pratt, Carrie. 1988. Computers as a progressive tool. In *Computers for transformation in education*, Cyril Julie, Ashiek Manie, Dirk Meerkotter, Rod Prodgers, and Doug Reeler. Bellville, South Africa: Wyvern Publications.

Postman, Neil. 1992. *Technopoly: The surrender of culture to technology*. New York: Random House.

———. 1986. *Amusing ourselves to death: Public discourse in the age of show business*. New York: Random House.
Rorty, Richard. 1998. *Truth and progress: Philosophical papers*, Volume 3. New York: Cambridge University Press.
Sclove, Richard. 1995. *Democracy and technology*. New York: Guilford.
Sen, Amartya. 1992. *Inequality reexamined*. Cambridge, MA: Harvard University Press.
Smith, Merritt Roe, and Leo Marx, Eds. 1994. *Does technology drive history?* Cambridge, MA: MIT Press.
Symons, Julian. Ed. 1957. *Thomas Carlyle: Selected works, reminiscences and letters*. Cambridge, MA: Harvard University Press.
Toffler, Alvin, and Toffler, Heidi. 1995. *Creating a new civilization: the politics of the Third Wave*. Atlanta: Turner Publications.
Touraine, Alain. 1995. *Penser le sujet*. Paris: Librairie Artheme Fayard.
———. 1997. *What is democracy?* Translated by David Macey. Boulder, CO: Westview Press.
———. 1990. The idea of revolution. In *Theory, Culture & Society*, 7: 121-141.
Weber, Max. [1905] 1996. The Protestant ethic and the spirit of capitalism. In *Knowledge and postmodernism in historical perspective*, edited by Joyce Appleby, Elizabeth Covington, David Hoyt, Michael Latham, and Allison Sneider. New York: Routledge.
Weibe, Robert H. 1995. *Self-rule: A cultural history of American democracy*. Chicago: University of Chicago Press.
Williams, Raymond. 1976. *Keywords*. New York: Oxford University Press.
Williams, Rosalind. 1994. The political and feminist dimensions of technological determinism. In *Does technology drive history?*, edited by Merritt Roe Smith and Leo Marx. Cambridge, MA: MIT Press.

Chapter Two

Breaking the Silence: Fax Transmissions and the Movement for Democracy in Malawi

Kedmon N. Hungwe

This chapter describes the political changes that occurred in the central African country of Malawi beginning in 1992. At that time Malawi began making the political transition from a single-party dictatorship ruled by the powerful Dr. Hastings Kamuzu Banda to multiparty democracy. Political change in Malawi was achieved through protest; for example, faxes and underground publications were used to break through repression and the severe censorship of information imposed by the state in thirty years of rule. With time, as the political climate became more favorable, these activities matured into newspaper publishing. This chapter focuses on the use of faxes and underground publications as a form of political resistance in Malawi.

The use of faxes in Malawian politics has been a subject of discussion in other publications (Lwanda 1995). This chapter draws on earlier published accounts as well as first hand experiences of individuals who participated in faxing. Their accounts broaden our understanding of an important phase in the history of Malawi. Political changes in Malawi, however, did not occur in isolation but were related to other global trends, particularly to changes in Eastern Europe.

Global Context for Political Change

The political changes in the Soviet Union associated with Gorbachev's glasnost and perestroika were a catalyst for political change in Africa generally, including Malawi. They considerably eased East-West tensions and made superpower rivalry less important as a geopolitical consideration. In this new political environment, human rights issues became ascendant in diplomacy and pressures for political change increased on dictatorships such as the one in Malawi. Furthermore, advocates for human rights, such as Amnesty International, Africa Watch, and opposition political groups had a more sympathetic hearing when they made the case for sanctioning governments historically allied to the West for their abuse of human rights. A wave of political change began to move through Africa, affecting countries such as Kenya, Zambia, South Africa, Namibia, and Malawi. These countries, among others, witnessed multiparty politics for the first time in decades.

Multiparty politics have not necessarily been a panacea for African political problems. Some dictatorships have manipulated the process to legitimize their corrupt regimes, Kenya being a case in point. It's probably too early to make a judgment on the long-term prospects for the fledgling democracy of Malawi. However, profound changes did occur beginning in 1992 when the people of Malawi found the courage to protest their long repression and succeeded in ousting Dr. Kamuzu Banda and his Malawi Congress Party (MCP) from power through popular elections. I begin the account of these changes with a review of the history of Malawi and the political conditions which preceded the changes.

Background of Malawian Politics

Malawi is a landlocked country located in Southern Africa (Figure 2.1). It is slightly larger than the state of Pennsylvania and has a population of about ten million people. Malawi ranks among the world's least developed countries. Its economy is predominantly agricultural with about 90% of the population living in rural areas. Agriculture accounts for 40% of the GDP and 90% of export revenues. The cash crops are tobacco, sugar cane, cotton, tea, and corn.

The country has had a long-standing dependence on economic assistance from individual donor nations, the World Bank, and the International Monetary Fund (IMF).

Malawi was colonized by the British in 1890 and became an independent nation on July 6, 1964. Hastings Kamuzu Banda, a physician trained in the United States and Great Britain, was its first postcolonial leader. He was to become the undisputed leader of the country for the next thirty years, having declared himself the "Life President" of Malawi. Under his thirty-year rule, the history of Malawi in many ways became the history of Banda. The emergence of Banda was a phenomenon of the peculiarity of Malawi politics in the late 1950s.

Colonial Malawi Politics and the Emergence of Banda

Hastings Kamuzu Banda was born in Malawi, but there are some uncertainties about his date of birth. However, a reasonable estimate places his year of birth at about 1898. What is known for certain is that he left Malawi for Southern Rhodesia (now Zimbabwe) in 1915, following an established Malawian tradition of migrating southward in search of work. He spent some years in Southern Rhodesia before moving to South Africa, where he worked underground as a miner. He had a keen interest in education, and when an opportunity to go to the United States for further study presented itself, he took it.

Banda was educated in the United States in philosophy, history, and medicine. He later moved to the United Kingdom, where he gained more specialized qualifications in medicine before setting up a private medical practice there. In the late 1950s, he moved to the newly independent country of Ghana. It was from there that he returned to Malawi in 1958, having lived for four decades outside his own country.

Banda's Return

Banda returned to Malawi by invitation from nationalist politicians who were struggling for the independence of Malawi from British rule. The leadership of the nationalistic movement was college edu-

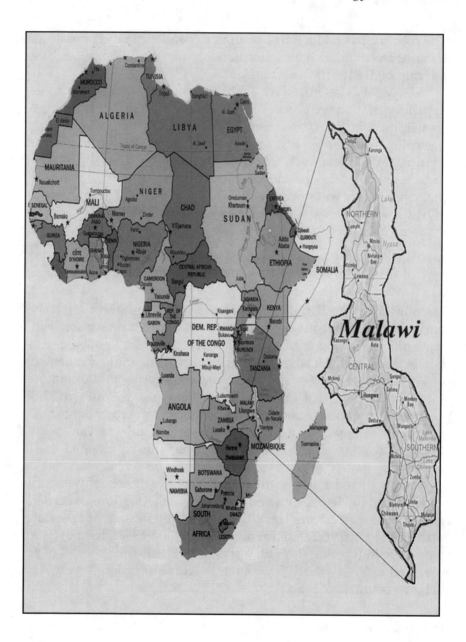

Figure 2.1. Malawi and Surrounding Countries

cated and relatively young, and was having difficulties forming a united front against the colonialists. The all too important rural constituency led by traditional chiefs had difficulties accepting the leadership of this "young" group. Without unity in the nationalist ranks, it was proving difficult to negotiate for independence from the British. Indeed, the British government had taken steps to entrench colonial rule of Malawi by setting up a federation of Malawi, Southern Rhodesia, and Northern Rhodesia (now Zambia). The federation effectively extended the harsh brand of Southern Rhodesian white rule to Malawi, a development that Malawians were determined to resist. However, they were plagued by divisions among themselves and were having difficulties creating a common political front. It was in this political climate that Banda was invited to return to Malawi to lead the nationalist movement. Kanyama Chiume, who was part of the "young" nationalist leadership that invited Banda, reflected on this period:

> By 1957 it was quite obvious that to preserve the desired unity and strong stand of the Party a leader who combined both age (thus acceptable to the old) and who at the same time spoke the political language of the imperialists and colonialists was desperately needed. The movement was threatened with total collapse through factionalism. Such a leader was to act as a bridge between the young and the old as they embarked upon an even more intensified united struggle against the Central African Federation and for the independence of Malawi. The old man picked for this role was Hastings Kamuzu Banda, then living in tightly concealed and unpublicised disgrace in Ghana. (Chiume 1992, 23)

Chiume's negative comments about Banda can hardly be taken to reflect his views back in 1959. His reservations about Banda appear to have developed after Banda had assumed leadership of the Malawi Congress Party.

Coalescing of Colonial Opposition

The letter inviting Banda to take over the leadership of the nationalist group was written by Masauko Chipembere. The movement offered Banda a powerful leadership position. Members of this group believed that the people of Malawi needed an inspirational figure: "Human nature is such that it needs a kind of hero-worship if a political struggle is to succeed" (Chiume 1992, 23). Indeed, the nationalist movement went out of its way to build up the Messianic image of Banda as a way of advancing its political goals, publicly referring to him as the Messiah. As Chipembere put it, "publicity of this sort could be used with advantage; it would cause great excitement and should precipitate almost a revolution in political thought" (Lwanda 1995, 56). Reflecting on this initiative decades later, Chiume concluded, "Little did the dedicated patriotic youth realize that they were making a tragic historical mistake for hardly a few years hereafter, Banda turned into a Frankenstein" (23).

Banda was successful in forging a united front against the British and negotiating for the independence of Malawi. In July 1964, Malawi attained independence. However, even as Banda was uniting the country against the British, he was also becoming increasingly alienated from key members of his party, and his colleagues were growing uneasy about his dictatorial tendencies.

Beginning of Banda's Repressive Rule

In September 1964, two months after the formation of the first postcolonial Malawi government, there was a government crisis when Banda's cabinet challenged his autocratic style of governing and expressed reservations about key elements of his domestic and foreign policies. Banda's response was to fire the leadership of the cabinet. Other colleagues resigned in sympathy. The sacked ministers included the group that had initially written to Banda inviting him to return to Malawi to lead the nationalist movement: Henry Chipembere, Kanyama Chiume, Augustine Bwanausi, and Dunduzu Chisiza. Banda made it clear that his word was to be law in Malawi. He derogatorily referred to his cabinet ministers as "boys" and was contemptuous of any form of dissent. In many ways the following statement, from one of his speeches, summarized his perspective:

> This kind of thing when a leader says this, but somebody else says that is not the Malawi system. The Malawi system, the Malawi

style is that Kamuzu says it and it is just that, and then it is finished, whether anyone likes it or not, that is how it is going to be here. No nonsense. No nonsense. You can't have everybody deciding what to do. (Chiume 1992, 27)

Banda's hand was strengthened by his ability to gain the support of the West, particularly Great Britain and the United States. He alleged that opposition politicians were aligned with communist China. His pro-South African policies also won him some allies, certainly in South Africa, which responded with financial aid, and also in some Western capitals. Malawi was the only African country to establish diplomatic relations with the apartheid government of South Africa. For that policy, it became a pariah state in Africa, and several African countries were willing to allow the political activities of exiled Malawian groups on their soil. In the aftermath of the cabinet crisis, the opposition had gone underground and fled the country. Banda responded by dispatching agents to intimidate and assassinate opponents based in other countries. Among those killed were Attati Mpakati, the leader of the Socialist League of Malawi, who was killed in Harare, Zimbabwe, in March 1983, and Alkwapatira Mhango, a journalist who was killed in Lusaka, Zambia, in October 1989 (Africa Watch 1990). One of the key instruments of Banda's repression was the Malawi Young Pioneers (MYP), a movement of young people originally formed to promote development. He turned it into a private paramilitary force, which terrorized and murdered people. Banda declared:

> The Young Pioneers can not be arrested by any policeman without my consent. ... If a Young Pioneer arrests anybody ... and brings them to the police station, the police officer in charge of the station must not release them. ... If he does release them, he is committing a crime. (Chiume 1992, 51)

In addition to the MYP, other important tools of Banda's control were draconian censorship laws.

Censorship and Control under Banda

Banda had total control over the press in Malawi. There were two state-controlled newspapers in the country, the *Daily Times* and a weekly paper, the *Malawi News*. The most important channel for information across the country was the government-controlled radio station, Malawi Broadcasting Corporation. Television was banned as a pernicious influence, and it was an offense to publish news about Malawi except through officially sanctioned channels. The Malawi Censorship Board monitored the flow of information and banned numerous books, films, and stage plays. Among the many banned books were those by George Orwell, James Baldwin, and Wole Soyinka. Telephone calls, particularly international calls, were monitored, and mail was tampered with. With rare exceptions, no foreign journalists were allowed to reside in Malawi.

Banda went beyond the control of information to the legal enforcement of quasi-Victorian dress codes. The mini-skirt was banned, as was long hair for men. The song "Cecilia" by U.S. recording artists Simon and Garfunkel was banned because it was deemed offensive to Banda and his longtime companion Cecilia Kadzamira. The lyrics, which included the words, "Cecilia, you're breaking my heart," were not to Banda's liking, since his relationship with Ms. Kadzamira was troubled. The story demonstrates the total and whimsical power that Banda accumulated over the years. Thousands of Malawians were imprisoned for years without trial for frivolous offenses or for unfounded suspicions of antigovernment activities. Some died in atrocious prison conditions (Africa Watch 1990; Lwanda 1995; Lawyers Committee for Human Rights 1992).

Resurgence of Malawian Political Resistance

By the late 1980s, Banda's health was failing, the economy was in serious decline; and there appeared the beginnings of unprecedented political changes in Malawi and other parts of the world. Nelson Mandela was released from prison in February 1990, and the Soviet Union was disintegrating. The first signs of overt opposition inside Malawi occurred in late 1988, when a group of students at the University of Malawi published a pamphlet, the *Chirunga Newsletter*, critical of government policies. For that action, four of them were expelled from school in January 1989; the expulsion led to public student demonstrations. Eventually, some of the students had to leave the country for their own safety. The next important publication came some years later from the Catholic church in the form of a pastoral letter. It generated unprecedented public protests against the government.

Pastoral letter and public protest. On March 8, 1992, in Catholic churches throughout Malawi, a pastoral letter was read that was the most significant challenge to the Malawi regime to be openly published and disseminated in decades. The letter, entitled "Living Our Faith," attracted attention beyond Catholic congregations, and its message resonated with the desires and feelings of Malawians from all walks of life (Drummond 1993; Lwanda 1995). The reading of the letter is widely regarded as a watershed event that triggered political change in Malawi (Newell 1995). The letter[1] began:

> As Pope Paul VI says: "the Church is certainly not willing to restrict her action only to the religious field and disassociate herself from man's temporal problems". In this context we joyfully acclaim the progress which has taken place in our country, thanks in great part to the climate of peace and stability which we enjoy. We would, however, fail in our role as religious leaders if we kept silent on areas of concern.

One of the key areas of concern was the suppression of dissent in a country where Banda's word was law.

> No one person can claim to have a monopoly of truth and wisdom. No individual—or group of individuals—can pretend to have all the resources needed to guarantee the progress of a nation. ... The contribution of the most humble members is often necessary for the good running of a group.

The letter addressed the issues of human rights and censorship.

> Academic freedom is seriously restricted; exposing injustices can be considered a betrayal; revealing some evils of our society is seen as slandering the country; monopoly of mass media and censor-

ship prevent the expression of dissenting views; some people have paid dearly for their political opinions; access to public places like markets, hospitals, bus depots, etc., is frequently denied to those who cannot produce a party card; forced donations have become a way of life.

The bishops expressed the belief that ordinary Malawians were anxious for change. People at all levels were urged to respond to the message.

We urgently call each one of you to respond to this state of affairs and work towards a change of climate. Participation in the life of the country is not only a right; it is also a duty that each Christian should be proud to assume and exercise responsibly.

After the publication of the pastoral letter, the bishops were summoned to meet the Inspector General of Police on March 10, 1992, and were questioned for eight hours. The letter was declared a seditious document and banned. Possession and distribution of it became an imprisonable offense. However, countrywide, the spell had been broken and the letter stirred unprecedented demands for political change and better economic conditions in this impoverished nation. There were spontaneous demonstrations for the first time in the history of postcolonial Malawi. The changing climate was reflected in a Department of State advisory to United States citizens:

The Department of State advises U.S. citizens to exercise caution while traveling in Malawi, as spontaneous civil disturbances are liable to occur. Although no specific threats have been directed against American citizens, recent strikes and demonstrations in the city of Blantyre in the southern region and the capital city of Lilongwe in the central region have become violent at times. U.S. citizens should avoid any areas in cities where large crowds are gathering. (1992)

Within days of the publication of the pastoral letter, its distribution and discussion became a major act of political resistance in Malawi. It was illegally photocopied and faxed within Malawi and overseas.

The government was alarmed, and at an extraordinary meeting of the ruling Malawi Congress Party, there were calls for the assassination of the bishops. Eventually Monsignor J. Roche, an Irish expatriate and signatory of the letter, was expelled in a retaliatory measure. However,

> by the time, for example, the Bishop's letter was being discussed by the MCP executive, not only had it been faxed to the BBC and the rest of the world, but the 16,000 print run document had been widely photocopied. (Lwanda 1995, 271)

The rapidly changing situation inside Malawi created new opportunities for opposition groups, both within and outside the country.

Resurgence of Organized Political Opposition

Opposition politicians who had fled Malawi during Banda's rule were mostly concentrated in countries such as Zambia, Zimbabwe, and Tanzania. Over the years, these parties had been weakened by a number of factors: they were plagued by divisions among their ranks, which made it difficult to form a united front against Banda; their support from host governments sometimes wavered depending on trends in regional politics; and Banda had agents in foreign capitals who hunted and assassinated his opponents.

In 1992, Malawi opposition politics witnessed a resurgence in Zambia. In October of the previous year, Zambia had made an historic transition to multiparty politics, and the political climate there was most favorable for Malawi opposition politics. It was in this new climate that an historic meeting of the fractured opposition in exile was held in Lusaka in April 1992. Coming in the aftermath of the bishops' letter and the rising political consciousness in Malawi, this was the first conference in decades to bring together opposition figures from inside and outside Malawi. One of the individuals who attended from inside Malawi was Chakufwa Chihana. The meeting led to the formation of the United Front for Multiparty Democracy, which later evolved into Alliance for Democracy (AFORD), led by Chihana. The event was well covered by the international media, particularly the BBC. The purpose of the alliance was to spearhead democratic change in Malawi.

Chihana's arrest. At the end of the Lusaka meeting, Chihana returned to Malawi to organize political activities within the country and was arrested on arrival at Lilongwe airport. At the time of his arrest, he was attempting to read a statement drafted by the opposition, so the statement and news of his arrest were faxed around Malawi and the world by opposition groups. He became a symbol for the protest movement. The meeting in Lusaka and the subsequent arrest of Chihana were important events in maintaining an international focus on Malawi. The opposition enjoyed considerable success with the Lusaka meeting and its handling of the Chihana episode. That experience demonstrated the importance of a coordinated information campaign to sustain the opposition movement inside Malawi.

The Information Campaign

One of the key resource persons in the development of the information campaign was Mike Hall.[2] His interest in Malawi politics was a natural outcome of his having been born and raised in that country. He was, however, a British citizen. In October 1988, he had applied for a permit to work as a BBC news correspondent based in Malaya: "They allowed me and my partner at the time to set up as correspondents, which was unheard of." Hall had friends in government, which probably helped him to win approval. The government was also anxious for publicity to attract foreign aid, since it needed aid to cope with a huge influx of refugees from the civil war in Mozambique. Hall and his partner worked in Malawi for a year before being expelled: "We were given 24 hours to leave." He had filed a story about the forced purchase of Malawi Congress Party cards.

> In that story I said that even pregnant women were sometimes harassed by the Malawi Young Pioneers to buy party cards for themselves and their unborn baby. Which happened, you know. They used to stop women from getting on buses and then say, where is your card, and where is the card for your child when she was

> obviously pregnant. The Malawi government, particularly, I think
> Madam Kadzamira, was very upset about it.

Kadzamira was not only Banda's longtime companion, but the leader of the national women's organization, Chituko Cha Amai Mu Malawi (CCAM). By 1992, when Banda's health began to fail, she was believed to have assumed a great deal of power and, with her uncle John Tembo, was in charge of the day-to-day functions of government (Africa Watch 1990).

After his expulsion from Malawi, Hall moved to Lusaka, in neighboring Zambia, where he continued his journalistic work. There he was involved in setting up an independent Zambian paper, *The Post*.

> It was just before the multi-party elections took place in Zambia.
> The first publication came out in July 1991. And so I was involved
> in that. I was a shareholder. And then I became editor in 1992.

Lusaka was a center of Malawi opposition politics; over time Hall became an active participant in those activities. He was a prime mover in the establishment of an opposition newspaper, the *Malawi Democrat*. His relationship with the Zambian newspaper the *Post* was critical.

> We used the facility of the *Post* to produce the *Malawi Democrat*, to
> begin with. So, in a sense, the *Post* was very, very useful to enable
> us to have the facilities, the desktop publishing and so on, to pro-
> duce the *Democrat*.

Use of Fax Technology

From the beginning, the publication of the *Malawi Democrat* was coordinated with faxing to Malawi. The paper was funded by a Malawian businessman based in Zambia and some Asian business people in Britain. It was smuggled into Malawi.

> We had distributors who would come to Lusaka, and buy second hand clothes in Lusaka, and they would take a bundle of copies of the newspaper, take them into Malawi, and sell the newspaper as well. We didn't care if it was sold or it was free as long as it was getting into the country. But apparently it used to sell like hot cakes. But we never saw the money. Our distributors would take the risk of getting caught so they kept the money. That was part of the deal.

The paper was published only fortnightly, which was not frequent enough to keep up with the rapid political developments in Malawi. That is why faxes became important.

> With the faxes you were able to respond the same day. Maybe, you know, someone is appearing, Chihana is appearing in court tomorrow. We would send out faxes saying Chihana is appearing in court tomorrow, go along and campaign for democracy. So the faxing, in a way, was much more immediate than the newspaper.

The addition of computer technology. The fax and newspaper operation expanded over time, and eventually an office was set up independently of the *Post*. The expanded operation continued to be funded by business people in Zambia and Britain and to a lesser extent by European aid groups who donated equipment. They had an Apple Macintosh computer, a photocopier, a fax machine, and telephone lines. Two prominent opposition figures, Frank Mayinga and Mapopa Chipeta, were involved with the newspaper and faxing. "And in terms of the faxing it was very much Frank Mayinga, who was behind it," said Hall. When the political climate in Malawi permitted it, the offices of the *Malawi Democrat* moved to Lilongwe. Eventually, Frank Mayinga became a cabinet minister in the new government.

Hall believed that faxes had "some effect" on political change, particularly in reinforcing "what was going on Malawi." In other words, if faxing

> had happened earlier, in another time, without the other conditions that were present, I can't think it would have had any effect.

Because Malawians had been exposed over the years to propaganda from exiles. But the conditions just didn't exist inside the country for it to foster any kind of meaningful movement. But with the bishops' letter, and a number of other factors, there was a climate already taking place there and people said, yeah, this is something else we can do. It was a kind of a novel thing, a novel form of protest.

Protesting with faxes inside Malawi. Inside Malawi, there was a great deal of interest in reading and distributing faxes. It became part of the protest activity, together with street demonstrations and work stoppages. These activities were fraught with risk, but people were determined to press for change. One person who participated in the distribution of banned documents inside Malawi was Dede Kamkondo, an author of literary fiction based at Banda College. His personal experiences of repression under the Banda regime motivated him: "I really wanted change," he remarked.[3] He had witnessed the arrest and harassment of prominent Malawian academics and authors such as the renowned author Jack Mapanje, who had been incarcerated for unspecified reasons. Kamkondo himself had been intimidated and harassed by government officials since his freshman year of college when he published his first work of fiction. The piece, entitled *The Vacant Seat*, had aroused the wrath of censors because they interpreted it as a threat to Kamuzu Banda. *The Vacant Seat* was the story of a struggle for succession to a chieftainship which ended in murder when the rightful heir was slain by his brother. Kamkondo could not understand how such a plot could be construed as subversive.

On graduating from college, Kamkondo initially worked as a reporter for the state-controlled paper, the *Daily Times*. Again, he ran into difficulties with what he thought were routine news stories. Concerned about his safety, he resigned. The effect of these experiences was to make him keen to see political changes in Malawi. When the chance came to participate in the protest movement which began in 1992, he readily joined. He was part of an underground group at his college that distributed and faxed documents within and outside Malawi. One source of their stories were banned publications. In addition to faxing factual information, they also generated false information designed to destabilize the regime. One message might, for example, allege that Banda had been responsible for a murder, even when there was no evidence or proof of that. The fax movement was therefore, in many ways, a propaganda war against the government.

Faxes coming from inside and outside Malawi were targeted to all fax machines that the opposition could access. Many of the targeted machines were in government offices, including the police, and some were in private companies. Government departments and civil servants were special targets for two reasons: Their support was considered critical for the success of the pro-democracy movement, and government agencies had the most elaborate network of fax machines across the different regions of the country.

The proliferation of fax machines within government offices in Malawi was due in part to international assistance programs. Fax machines became a part of the aid package from organizations such as the International Fund for Agricultural Development (IFAD) and the World Bank and were intended to facilitate communications. There was, for instance, a fax machine in every district agricultural office. Within the capital, Lilongwe, the main post office received hundreds of anonymous faxes from inside and outside Malawi. One person who saw the faxes was Rose Muwalo,[4] a post office employee responsible for handling faxes and telexes. Muwalo and her colleagues were under orders not to read or distribute anonymous faxes. "We tore them up," said Muwalo. "I did not even tell my husband about the faxes." She worried that her husband would share the faxes with his friends and "I could be revealed as the source," she said. Muwalo had reason to be concerned. During Banda's rule, a security officer was permanently stationed at the post office, disguised as a postal worker. He checked suspicious mail and spied on employees. A colleague of Muwalo was suspended for forwarding a fax to Chihana's office, even though the message had nothing to do with opposition politics. One of the documents which came through the post office fax machine was the bishops' pastoral letter.

> The bishops' letter was hot. People got rid of it. It was faxed around. Possession of it was a criminal offense. Because of the dangers of

distributing faxes, people ended up just throwing them out into the street, for people to pick up and read.

Other faxes coming through were articles critical of the regime from foreign newspapers such as the *New York Times, British Guardian,* or *Zimbabwe Herald.* Those articles would not normally be accessible to Malawians.

Another state-controlled agency which received numerous anonymous fax messages was the Electricity Supply Company (ESCO). Security officials raided the premises and detained all the employees in the computer section of the company. They were charged with possession and distribution of faxes. Muwalo's uncle was among those arrested.

The government raided several private companies as well. They targeted companies with international communication facilities such as the National Bank, where twenty employees were arrested (Lwanda 1995). Despite the harassment and intimidation, protests through faxes were unabated. As a result, Malawians were better informed and better able to keep the world abreast of developments in their country. On May 11, 1992, two months after protests began, Western donors suspended all nonhumanitarian aid to Malawi. On December 31 of the same year, Banda called for a referendum to decide on the political future of Malawi. That referendum was held on March 15, 1993, and led to Malawi's first-ever multiparty elections on May 17, 1994. The elections brought an end to the rule of Kamuzu Banda and his Malawi Congress Party.

From Faxes to Newspapers

Malawi experienced an important political transition in the early 1990s. Several factors within and outside Malawi were important in making that transition possible. The determination of Malawians to stand up to their government was a key factor in the transition. Faxes became an important tool in raising consciousness, mobilizing and coordinating resistance against the state, and keeping the world informed of changes in the country. They were an integral part of the struggle for human rights and democracy.

In post-Banda Malawi, some of the creative effort that had been associated with the fax movement was turned into newspaper publishing. In 1994 there were more than twenty newspapers, with a broad spectrum of political views. The new government has been generally tolerant of criticism. The old repressive laws have been repealed under a new constitution. There have, however, been some lapses. In one well-reported incident, the police detained and harassed an editor whose paper had published an old photograph of the new president in prison uniform, supposedly taken after his arrest for theft (U.S. State Department 1995). On the whole, there have been marked improvements in the political climate. It appears that the people of Malawi have reasserted their right to participate in the process of shaping their destiny.

Notes

1. The letter was signed by Archbishop J. Chiona, Bishop F. Mkhori, Bishop M. A. Chimole, Bishop A. Assolari, Bishop A. Chaingwera, Bishop G.M. Chisendera, Monsignor J. Roche, and Father Gamba. The full text of the pastoral letter is given in Lwanda (1995).

2. The information attributed to Mike Hall is based on a telephone interview with him on November 14, 1996. He was then living in Australia.

3. The information attributed to Dede Kamkondo is based on an interview with him on September 15, 1996. He was then on sabbatical leave in the United States.

4. Rose Muwalo is a pseudonym. She was interviewed on October 10, 1996, in the United States.

References

Africa Watch. 1990. *Where silence rules: The suppression of dissent in Malawi*. Washington, DC.
Chiume, K. 1992. *Banda's Malawi*. Lusaka, Zambia: Multimedia Publications.

Drummond, E. 1993. Malawi joins Africa's democracy march. *Africa News* 38(6):6-7.
Lawyers Committee for Human Rights. 1992. *Malawi: Ignoring the call for change.* New York.
Lwanda, J. L. 1995. *Kamuzu Banda of Malawi: A study in promise, power, and paralysis (Malawi under Dr. Banda) (1961-1993).* Glasgow, Scotland: Dudu Nsomba Publications.
Newell, J. 1995. "A moment of truth"? The church and political change in Malawi, 1992. *Journal of Modern African Studies* 33 (2): 243-262.
U.S. State Department. May 11, 1992. Travel Advisory-Malawi.
———. 1995. *Malawi Human Rights Practices,* 1994.

Chapter Three

Resistance and Cybercommunities: The Internet and the Free Burma Movement

Zarni

Within a relatively short period of time, Burmese dissidents in exile and their supporters around the world have transformed what had been an obscure democratic struggle in mainland Southeast Asia into one of the largest human rights campaigns in the world (Urschel et al. 1996; Barron 1998; Charles 1998).[1] In as many as twenty-eight different countries, Burmese dissidents and their supporters have made use of the Internet to form coalitions and share strategies in their efforts to weaken the grip of the military rulers of Burma. Specifically, they have taken advantage of the Internet to develop communities, to educate a global audience about the political situation in Burma, and to support Burma's democracy movement through economic activism and human rights advocacy.

For Burmese dissidents in exile, the Internet has provided a critical voice for people who are silenced in their own country. Although Burma's military junta tightly controls all media within the country, it cannot stop dissidents from reporting the junta's activities to the international community. Often unable to travel, Burmese dissidents who operate along Burma's border areas and in urban areas of

Burma's neighboring countries such as India, Thailand, and, to a lesser extent, Bangladesh, have nevertheless been able to create a critical link between their fellow dissidents in various countries and concerned citizens around the world through the Internet. These dissidents closely monitor the developments inside Burma in terms of human rights, the junta's troop movements, economic conditions, public opinion, politics, and foreign investment. A worldwide network of activists, organizers, and supporters of Burma's democracy movement use this information to alert the international media about potentially newsworthy events inside Burma, to update various sympathetic international organizations and governments around the world, to recruit individual supporters and members for the movement, and to chart various campaign strategies.

This chapter discusses the emergence of the Free Burma Movement, an international grassroots effort coordinated through the Internet. I focus specifically on the process of how these political communities have emerged and on the dynamics concerning Free Burma activists and organizations with diverse national, ideological, and geographical backgrounds. I conclude this essay with a discussion of how what I call a "Free Burma identity" is forged among those who are part of the Free Burma cybercommunity. Although I draw on some outside sources, these observations are based primarily on my ten-year long involvement in the struggle.

Formation of an Independent Burma and Rise of the Military Junta

Burma[2] is a Southeast Asian agrarian nation with some forty-five million multiethnic inhabitants. It borders Tibet, India, Bangladesh, Thailand, Laos, and China. In 1947 Burma regained its independence from Britain and adopted a Western-style parliamentary democracy. In 1962 the country's civilian government was overthrown in a military coup. From 1962 to 1988, the military leaders who formed the Burma Socialist Program Party (BSPP) government pursued an isolationist policy toward the world. Once a rich country, Burma was granted Least Developed Country status by the United Nations in 1987.

Suppression of Dissent

With a worsening economic situation and political oppression, Burma erupted into a series of pro-democracy popular revolts in 1988, which were violently suppressed by the ruling military junta. Consequently, about ten thousand Burmese students, civilians, and Buddhist monks fled to Indo-, Thai-, Bangladeshi, and Sino-Burmese border areas. Among these were democracy activists who joined with such armed ethnic minority groups as the Karen National Union, Kachin Independence Army, and the New Mon State Party to overthrow the military junta. Over several years there was a common realization among Burmese student revolutionaries that the prospect of restoring democracy in Burma through armed struggle was fading. Scores of students began to leave the border camps for urban areas in Thailand and India. Several hundred of them who passed minimum educational screening were brought to countries such as the United States, Norway, Australia, and Britain under various governmental sponsorships enabling those students to resume their studies interrupted by the 1988 uprisings and subsequent involvement in the armed struggle.

The Beginnings of an International Movement. As early as 1987, Burmese exiles in the United States established an international network of Burmese dissidents and politically minded professionals who were scattered around the world. The United States-based Committee for the Restoration of Democracy in Burma (CRDB) was the best-known political group, with chapters in Asia, Australia, Western Europe, and Canada. Students and other dissidents who fled after the 1988 crackdown formed the All Burma Students Democratic Front (ABSDF) and began making contacts with various international organizations and politically active Burmese exiles in other parts of the world. Soon after the establishment of the ABSDF in the Thai-Burmese border region, the CRDB began working with student revolutionaries who managed to retain close contact with those activists inside the country. In addition, nongovernmental organizations provided humanitarian aid to the refugees from Burma's ethnic minority communities who fled the Junta's military offensives and forced labor. Some humanitarian workers, especially those from industrialized nations such as the United States, played a crucial role in building bridges between these Burmese students-cum-revolutionaries and the church-, community-, and university-based organizations in Western nations.³

These individuals and organizations worked on various political projects including information gathering and dissemination; networking; provision of humanitarian assistance; arranging speaking tours of ABSDF representatives in Western democratic countries; and visits by supporters, Burmese and non-Burmese, to ABSDF camps in Burma's border areas. Noteworthy is the fact that the focus of their work was primarily to solicit funds and other political support for the Burmese revolutionary groups on the borders. In addition, they organized demonstrations and candlelight vigils before Burmese diplomatic outposts in cities such as Tokyo, Canberra, Bangkok, New Delhi, New York, Washington, D.C., London, Bonn, and Ottawa to protest ongoing political oppression and human rights violations in Burma. They hosted international gatherings of Burmese dissidents in countries such as the United States and Japan and formed traditional hierarchy-based exile organizations. They also published bilingual (English and Burmese) newsletters.

Rising political repression. In Rangoon, the capital of Burma, the military government assumed the name, "State Law and Order Restoration Council" (SLORC), and changed the country's name from Burma to Myanmar in 1989. It placed most of the key leaders of the democracy movement in jail or under house arrest, including Aung San Suu Kyi, cofounder of the National League for Democracy (NLD) and winner, in 1991, of the Nobel Peace Prize. In 1990 the SLORC held multiparty elections. The National Unity Party, a reformed Burma Socialist Program Party backed by the junta, was resoundingly defeated by the NLD. The SLORC, however, annulled election results and refused to hand over power to the elected representatives. In the meantime, the SLORC opened the country to foreign investment in their efforts to march Burma along the "Burmese Way to Capitalism," adopting the Chinese mode of liberalizing the economy without relinquishing an authoritarian grip over the people.

Owing to the country's twenty-six-year-long self-imposed isolation, Burma's popular uprisings and the ensuing massacres in 1988 escaped the world's notice. Following the awarding of the Nobel

Peace Prize to Aung San Suu Kyi in 1991, Burma's freedom struggle began to attract global attention.

Arrival of the Internet and the Changing Face of the Free Burma Movement

As early as 1988, the Burmese dissident Coban Tun, a graduate student in physics at San Francisco State University, was using email to disseminate Burmese news. Later, area studies departments began using email to share information and news updates about Southeast Asian countries through newsgroups such as "soc.culture.thailand" or "soc.culture.indonesia." Tun typed in Burmese news from Western newspapers and posted these items to those lists because there was not as yet a separate newsgroup for Burma. Tun also visited ABSDF student camps along the Thai-Burmese borders and wrote brief eyewitness accounts of the conditions under which Burmese student revolutionaries lived. As a computer programmer, Tun also provided computer software and taught some student fighters computer basics. Almost all student revolutionaries lacked any knowledge of computers, let alone advanced information technology.[4]

In 1994 Douglas Steele, an American working with Burmese student revolutionaries in Thailand, received a grant from the New York-based Open Society Institute to establish BurmaNet, an online Burma news listserver. According to Steele, the primary purpose of BurmaNet was to facilitate communication among Burmese students and dissidents, who were scattered around the world. Later BurmaNet began sending out Burmese news to Burma watchers around the world who had access to the Internet.

Information Dissemination by Internet

Burmese dissidents in Bangkok, Chiang Mai, Tak, as well as Burmese border areas would collect information about political developments through the underground network inside Burma. Groups such as the ABSDF had created a research unit whose task was to collect and disseminate information and news to others who were by now connected via the Internet, mainly by email. Also, the Burma-focused nongovernmental organizations (NGOs) and lesser known human rights groups such as the Karen Human Rights Group became involved in information gathering and dissemination, all relying on the already established BurmaNet. Members of these groups would go in and out of refugee areas, and sometimes into ethnically controlled areas, to interview victims of human rights abuses by Burma's army and collect information about the environmental destruction caused by SLORC's economic policies. They would also document the movement of Burmese army troops and report on occasional clashes and battles between other ethnic "insurgent" forces and SLORC troops. They would then enter these data onto diskettes using laptop computers and express mail them to their colleagues in Bangkok for distribution on the Internet.

In addition to these democracy activists who operated along Thai- and Indian-Burmese border regions, non-Burmese activists officially went into Burma under various disguises for purposes of information gathering and dissemination. These undercover organizers and activists, mostly Westerners including Americans, Canadians, Dutch, British, and Norwegians, would clandestinely meet members of the National League for Democracy and other members of the Burmese underground in various Burmese cities. They became a vital link between the National League for Democracy leadership and the Burmese dissidents outside the country. They would also bring back pictures of economic projects where the use of prison and slave labor was widely known and pictures of new shantytowns with no running water or electricity where the SLORC forced urban dwellers to live. They would interview various citizens to gauge the political opinion toward the current SLORC. In addition, they would interview NLD leadership including Aung San Suu Kyi, and relay her messages to Burmese resistance communities in particular and the world community in general.

Occasionally, the NLD members and supporters inside Burma would fax to Burmese activist groups based in Bangkok information and news including the minutes of the NLD meetings; lists of democracy activists and organizers who were recently arrested by the military intelligence; the health conditions of the NLD members behind bars; power struggles within the SLORC itself; the list of foreign investors in Burma; and so on.

The Growth of Free Burma Communities in Cyberspace

Initially, several hundred subscribers of BurmaNet "consumed" Burmese information and news provided by BurmaNet. Subscription was automatic and there was no screening involved. Many of the subscribers were Burmese and Burma scholars, Burmese expatriates and dissidents of diverse backgrounds, non-Burmese people interested in Burma, Free Burma organizers in various countries with different professional backgrounds, non-Burmese NGOs, and members of certain diplomatic missions based in Rangoon. Some used the information and updates to alert foreign government ministries, international human rights organizations, the United Nations agencies, international media, and other interested groups and individuals. Some used the information for various grassroots campaigns aimed at stopping foreign investment in Burma. Some fed the information to lawmakers in sympathetic Western governments in order to persuade them to toughen their policies on Burma or take concrete actions against the SLORC. Furthermore, individuals and groups collected information from the BurmaNet as they contemplated or planned to enter what was in their minds a newly emerging market. And, when the SLORC intelligence began using the Internet for propaganda and intelligence purposes, they also latched on to BurmaNet.

After BurmaNet became a fairly well-known information and news source, more people began subscribing to the list and sharing their ideas, opinions, and information through the BurmaNet Conference, an unmoderated listserv where BurmaNet subscribers can post their messages freely. Among the initial users of BurmaNet Conference information were Burmese students who later left their fellow dissident students in Burma's border areas to continue studies in such countries as the United States, Canada, Australia, Norway, Japan, and England.

Gradually, a sense of identity developed among BurmaNet subscribers, who were pro-democracy activists and supporters. A key defining characteristic of this newly emerging community was its support for Aung San Suu Kyi's leadership. By the time BurmaNet was created, there was already a small number of individuals, primarily in the United States, Thailand, and Canada, who were advocating consumer boycotts and were engaged in shareholder, campus, and community activism against foreign investors with economic interests in Burma.[5]

As a symbolic protest against the SLORC, these Burma and Burmese activists avoided using the new names given by the SLORC to sites and to the country. For instance, both pro-democracy non-Burmese and Burmese alike would insist on using the anglicized words "Burma" and "Burmese" as opposed to new names "Myanmar" and "Myanmarese" given by the SLORC. Who used what terms became indicative of their political affiliation or attitude toward the NLD and the SLORC and thus either pro-NLD and pro-democracy or pro-SLORC and pro-business.

Both the inception of BurmaNet and the availability of new online news groups such as "soc.culture.burma" aided communication among Burmese dissidents and their allies in more technologically advanced countries. However, the use of online information and the new communicative medium created by the Internet was confined to the already existing pockets of Burmese dissidents and Burma watchers. Furthermore, these small pockets of advocacy groups failed to realize the potential of the Internet in their educational and economic activism work. Some of the Burma groups such as California-based Burma Project (USA), the Committee for the Restoration of Democracy in Burma and the Portland-based Project Maje were not online yet.

On the ground, Burmese dissidents from various countries, as well as non-Burmese human rights advocates and environmental activists, traveled on speaking tours to countries where there were already Burma support networks to garner grassroots support for the Free Burma Movement and to organize boycotts and protest rallies. These actions generated interest in and support for Burma's struggle.

In addition, there was no coordinating body of online Free Burma organizers that was able to build an effective grassroots base for economic activism. Nor was there a well-known grassroots organization with a capacity to absorb, retain, and properly channel newly generated interest and support. The lack of a national and/or global clearinghouse, communication hub, or other well-coordinated political group proved to be a formidable barrier to the forming of loosely aligned Free Burma activist and support communities. For these reasons, Burma campaigns internationally remained sporadic and uncoordinated and Burma remained an obscure political issue.

Establishment of the Free Burma Coalition

It was in the aforementioned Burmese political climate that the Free Burma Coalition (FBC) was created in September 1995. The Free Burma Coalition is one of the single largest Internet-based political community. It has active supporters and organizers at roughly 150 colleges and universities, thirty high schools, over one hundred community-based Burma support groups, and individual supporters in twenty-eight countries scattered in Asia, Australia, North America, and Europe. The two primary objectives of the FBC are (1) to end foreign investment in Burma under the current military dictatorship through economic activism, and (2) to build a genuinely grassroots international Free Burma movement in support of Burma's freedom struggle. It was inspired by and modeled after the worldwide antiapartheid movement that brought about the end of a racially oppressive system in South Africa. Since the creation of the FBC in September 1995, the Internet-based coalition has been very successful as it helped to bring the withdrawal of one dozen multinational corporations, most notably PepsiCo Inc., from Burma; persuaded twenty-one U.S. municipalities and the Commonwealth of Massachusetts to enact "Burma Free" business ordinances and laws; and contributed to the passage of a conditional economic sanctions bill which became law in the U.S. in October 1996. These actions helped to create a Free Burma consciousness among such political forces as the American labor movement, religious and women's groups, college and high school activists, lawmakers, college administrators, and even corporate executives and board members.

Needless to say, many historical and political factors, such as the release of Aung San Suu Kyi from her almost-six-year-long house arrest two months prior to the establishment of the FBC and the resultant political optimism and hope within the Burmese democracy movement, contributed to the successful organizing and growth of the Internet-coordinated Free Burma movement. My purpose here in describing at some length my role and self-assigned tasks is primarily to highlight how a particular conjunction of circumstances can open a space for effective grassroots organizing.

In September 1995, I founded the Internet-based Free Burma Coalition with the help of two University of Wisconsin graduate students, Todd Price and Alex Turner, who are U.S. and Scottish citizens respectively. Besides Alex and Todd, other Free Burma organizers contributed to building the grassroots campaign via the Internet.[6] Most of them were already established organizers in the United States and were among those largely responsible for the early Free Burma economic activism. There were also Burma supporters who made financial contributions to the FBC's initial recruitment activities.[7]

The FBC was and is registered as an official student organization at the University of Wisconsin, Madison. Organizing around an international event on October 27, 1995 publicized as the "International Day of Action for a Free Burma,"[8] the FBC organizers made good use of various media including human rights video-documentaries, the Hollywood film *Beyond Rangoon*, Free Burma print materials, and a Web site entirely devoted to the campaign activities in order to popularize Burma's freedom struggle.

In addition, the FBC distributed hundreds of professionally produced four-color campaign posters (see Figure 3.1) to be used as visual aids.[9] The campaign poster had a political message that reflected the aforementioned objectives of the FBC. Armed with nearly 100 copies of a twelve-minute-long Burma human rights video documentary and 500 copies of FBC posters, the FBC kicked off the Free Burma campaign at an international conference of the Student Environmental Action Coalition (SEAC) held on the campus of the University of North Carolina, Chapel Hill. Information packets, posters, and videotapes were distributed free to any organizer at the conference who signed up to participate in the planned International Day of Action for a Free Burma. Those who signed up were only asked to leave their email addresses so they could be updated about the human rights situation inside Burma.

Also at the conference were Burmese student leaders from the All Burma Students' Democratic Front, who were brought in from Thailand as speakers and panelists. After having heard first-hand accounts of Burmese students' struggle for democracy, American and other international students attending the SEAC conference were ready to help. When these American and international supporters learned of the FBC's plan for an international day of action, they readily signed up. At the end of the SEAC conference, UW-Madison graduate student John Peck organized a Pepsi-protest demonstration in front of a local Taco Bell restaurant near the University of North Carolina, Chapel Hill.[10] With the high-spirited Taco Bell protest, the Free Burma campaign had officially begun.

When students arrived back at their campuses many continued to talk via the Internet. Having felt empowered during the Chapel Hill demonstration, students were ready for local Free Burma protests as the date for the Free Burma Action Day drew near. In the meantime, the Free Burma Coalition email list was growing and scores of people from Asia, South Africa, Australia, and Europe were signing up for the action day. Those who were not present at the demonstration learned about it through the Internet. Activists shared narratives on the Internet, and the excitement and enthusiasm they generated proved contagious. The Internet created a virtual space where the flow of these activist energies was kept alive and rechanneled for ongoing political actions such as divestment and boycott campaigns.

Developing a Presence on the Net

On the Internet front, another UW-Madison graduate student, Alex Turner, developed a Web site (http://wicip.org/fbc) specifically for organizing purposes.[11] Although young and still not rich in its content, the site served the purpose at the time of the first Free Burma Action Day. It offered all the campaign materials needed for interested groups to participate in the "International Day of Action for a Free Burma." These materials included the Free Burma campaign poster, basic information about Burma's freedom struggle, pictures of Aung San Suu Kyi, texts of her speeches, a readymade flyer which could be downloaded and tailored to local needs, a list of groups that endorsed the Free Burma action, and so forth. Furthermore, the images from the first PepsiCo protest at UNC-Chapel Hill were immediately archived and their presence on the web was made known to FBC members. These images gave members a sense of belonging to, and of solidarity with, the movement as they reminded them-

Figure 3.1. Free Burma Action Day poster

selves what they had accomplished working together with a newly formed organization.

These newly created resources, in conjunction with the existing BurmaNet as a primary information reservoir, enabled key organizers of the Free Burma movement to mobilize support from activist communities with an access to the Internet and helped to retain the level of Free Burma consciousness among those already on board. This action was facilitated by the creation of an email listserv. As an official student organization, the FBC was able to secure an email list provided by the University of Wisconsin-Madison. The Free Burma list is private and usually moderated by me. As the sole list owner, I have the password to control the list. Subscriptions are handled manually to screen out potential users who would disrupt the group dynamic and to prevent possible infiltration by SLORC.

One-person ownership of the list obviously has its own problems. It could be argued that a moderated list is less democratic than an unmoderated one. To minimize possible inconsistencies in selecting postings from various groups and organizers, a set of criteria was developed and publicized on the list. As the moderator-editor of the list, I spent between twelve and fifteen hours a day working on the Net, with few breaks for the first several months. I began publishing daily editorials which were invariably upbeat in tone and carried an insider's readings of Burmese politics, culture, and economy. For a little over a year I handled inquiries, responded to every new recruit, created editorials, edited various daily postings, and wrote occasional action alerts. After a period of time I developed a personal rapport and cyber-camaraderie with almost all members and subscribers on the list.

Diversity in the movement. As a result of these private Internet communications early in my organizing efforts, I had an opportunity to find out that our supporters and allies from many different communities brought widely diverse political values, perspectives, beliefs, professional experiences, cultural traditions, religious orientations, and ideologies to the Free Burma Movement. FBC members and supporters ranged from left-leaning campus organizers who wanted to use Free Burma campaign as a political vehicle for ridding the universities of corporate interests and influences, to stock brokers and investment managers. As a Burmese, I felt it crucial to maximize the contributions these individuals could potentially make while keeping them from getting into serious conflicts on the Internet based on their seemingly irreconcilable political differences. While such diversity might prove to be a strength and value for the grassroots campaign, it might also create a tension-filled situation in cyberspace. A number of environmental and cultural factors later proved to be of great value in dealing with this situation.

My graduate education in cultural politics provided me with the kind of understanding and vocabulary vital to the creation of a new community, its sense of political identity, and a common political agenda. Being a Burmese in exile worked in my favor in terms of political legitimacy and authenticating my political voice. The institutional setting (namely, academia) in which I had been operating also lent credibility to my political work. I didn't fail to see the advantages of having a legitimate place in one of the premiere U.S. institutions with a fine reputation for both its education and its progressive tradition. One cannot fail to notice this institutional base from the coalition's mailing address: FBC c/o Department of Curriculum and Instruction, The University of Wisconsin, Madison, WI 53706. The mission statement clearly introduced the leadership of the movement—Aung San Suu Kyi and the National League for Democracy—and what role we as Free Burma organizers were to play. All of these factors lent credibility, enabled me to function across boundaries and precluded, to a considerable extent, internecine fights for the leadership. In addition, as the moderator of the main communicative medium, the Free Burma list, I made serious efforts to communicate to all members of the FBC in a political language using a set of vocabularies with which most could agree. Following is an example in which I paid close attention to the intricacies and the power of rhetoric.

Creating a Free Burma identity. In the daily editorial notes, endearing terms such as "fellow freedom fighters," "dear fellow spiders," or more serious terms such as "freedom struggle," "democracy movement," "working toward a Free Burma," rather than the more generic terms and phrases such as "campaign," "dear activists," "dear all," "dear organizers," were carefully chosen. The Ethiopian proverb, "When spiders unite they can tie down a lion," was borrowed from an email signature file and subsequently used as the

FBC motto. The Internet organizers and supporters of the Free Burma movement began to see the relevance of the proverb that creates a mental imagery of the seemingly weak defeating the powerful, a proverb with a strong political message—what Vaclav Havel aptly terms the "power of the powerless." Over a period of a year and half of almost daily bombardment with the word "fellow spiders," Free Burma activists from diverse political and professional backgrounds began calling themselves "spiders." Even the *Economist* talked about the appropriateness of the use of this motto for a grassroots movement such as ours (Long 1996). EuroBurmaNet, a site developed for and devoted to the Internet-based Free Burma movement, even created a section "Spiders at Work." This spider imagery and vocabulary became widely known among Internet-based organizers around the world.

These efforts in cyberspace aimed at creating a new sense of political identity and a community—both virtual and real— came to be known as the "Free Burma Community." The political consciousness built from the ground up was recognized, celebrated, and advanced when the leader of Burma's democracy movement Aung San Suu Kyi, at the FBC's request, specifically endorsed an international two-day fast for a Free Burma in October 1996 using the vocabulary already familiar to the Free Burma Netters.

Conclusion

The emergence and growth of a given community, its identity and sense of community, are a product of constant processes, few of which can be predetermined or carefully designed. To be sure, building a community, or more specifically, a collective political identity of those loosely in association with one another in cyberspace requires a great deal more than the contributions and effort of one person. As mentioned earlier, bringing all the key organizers from various countries on board, working closely with them, and quietly consulting them through private Internet communications are but a few elements that can be recognized in successful democracy campaigns.

The new communication technologies have greatly aided the rapid growth of a global Free Burma community both in cyberspace and real space. Since they mirror the actual historical and material conditions from which organizers participate, their virtual space reflects the contestations of the physical space of the movement. This said, the Internet nonetheless provides a rich new space for democratic activists to join in solidarity and plan resistive activities, where many small spiders can spin a powerful web.

Notes

1. Since its inception in September 1995, the Free Burma Coalition has generated nearly 200 articles in the international print media including the *New York Times*, the *Guardian*, the *Washington Post*, the *Japan Times*, *Los Angeles Times*, and *Chicago Tribune*.

2. The current military government of Burma changed the name of the country to Myanmar in 1989. Although the new name is sometimes used in news accounts, Burma democracy advocates refuse to use it.

3. Max Ediger, a Thailand-based humanitarian worker with strong ties to an Oklahoma church community, brought Moe Kyaw, the first representative from the ABSDF, on a speaking tour in the United States. Various branches of the Burmese dissident-in-exile network, CRDB, and a handful of politically minded Burmese students in the United States provided assistance for the tour.

4. Computer science as a field of study in Burma's higher education was introduced only a few years prior to the fall of the Burma Socialist Program Party government. Even then, fewer than 100 students were admitted to the computer science program at Rangoon University, the only place in the country where computer education was provided.

5. Edith Mirante of Portland, Oregon based Project Maje, Don Erickson of the Chicago-based SYNOPSIS, and Pam Wellner of the Berkeley-International Rivers Network are among the non-Burmese organizers. David Wolfberg of the Rainforest Action Network and Reid Cooper of Ontario Public Interest Research Group at Carleton University in Canada were the pioneers in North American Burma grassroots activism.

6. They were Professor U Kyaw Win of the Committee for Restoration of Democracy in Burma, John Peck of UW Greens at the University of Wisconsin-Madison, Larry Dohrs of Seattle Campaign for a Free Burma, Simon Billenness of the Boston-based Coalition for Corporate Withdrawal, David Wolfberg of Los Angeles Campaign for a Free Burma, Christina Fink of the Bangkok-based BurmaNet, Burmese dissident student Tun Myint of Indi

ana University at Bloomington, Nick Thompson of Student Environmental Action Coalition (SEAC) at Stanford, Marco Simons and Adam Richard of Burma Action Group at Harvard, Ohmar Khin of the Washington, D.C.-based Refugees International, Yuki Kidokoro of the UCLA SEAC group, environmental and Burma activists such as Mike Ewall, Brad Simpson, and Carwil James at Northwestern University, Brian Lipsett and Tony North at Pennsylvania State, and Linda Kwun of SEAC at the University of Illinois at Urbana-Champaign.

7. The individuals who made initial financial contributions to the FBC included the Santa Fe-based photojournalist Mimi Forsyth, Leslie Kean of the Burma Project/USA, the exiled National Coalition Government of the Union of Burma, Attorney Harry Salzberg of the Madison-based Equal Justice Foundation, Don Erickson of Synopsis in Chicago, and London-based travel writer Nicholas Greenwood.

8. See Tyson, A. S. 1995. Political activism on campus takes on a cyberspace twist. the *Christian Science Monitor,* October 31.

9. As education students, Todd Price and I were keenly aware of the importance of visual aids in human communications, be it teaching, learning, or organizing. This led us to work on the poster project with the help of Professor Elizabeth Ellsworth of the School of Education at UW-Madison.

10. At the time, Taco Bell was a subsidiary of PepsiCo, Inc.

11. There is a cluster of Web sites of varying degrees of use for the Free Burma campaign. For a good starting point, visit http://freeburma.org.

References

Barron, S. 1998. Burma's busy Networker. *The Nation,* 5 January.
Charles, D. 1998. Myanmar fighters swap guns for modems. *The Philippines Star,* 13 March.
Fink, C. 1998. Burma: constructive engagement in cyberspace? *Cultural Survival Quarterly.* 21(4): 29-33.
Long, Simon. 1996. Arachnophilia. *The Economist,* August.
Steele, D. 1997. Internet activism. Panel discussion at the first Free Burma Coalition Conference, at American University, Washington, DC. 1 January – 2 February.

Tyson, A. S. 1995. Political activism on campus takes on a cyberspace twist. *Christian Science Monitor,* October 31.

Urschel, J., et al. 1996. College cry: "Free Burma" activists make inroads with U.S. companies. *USA Today,* 29 April.

Chapter Four

Old Technology in New Contexts: Print Media and Russian Education

Stephen T. Kerr

In the West, discussions about educational change and school reform often feature technology as a central element. World Wide Web sites as bases for curricular innovation, distance education as a method to reach underserved populations, multimedia instructional software as a path to improved learning, virtual reality as the key to new modes of cognitive development—all these and more regularly appear as central tenets of the American vision of the future of education. Similarly, teacher education and the continued professional development of educators are typically discussed in terms of new kinds of electronic networks and new forms of communication. High tech is in; print, to the extent that it figures in these visions at all, is seen as something increasingly outmoded, a medium to be abandoned, a technology with a weak, slow, and imprecise impact.

Yet, in other parts of the world, print is still the medium of choice. Often this is so because there is simply nothing else available. In spite of our hopeful predictions, much of the world still does not have easy access to computers, multimedia systems, or the Internet.

And even if some of these resources are available, their configuration may be far from what we think of as minimally acceptable. It is more than a bit disingenuous to encourage the use of the World Wide Web as a curricular resource when the only available telecommunications connection is via a 300-baud modem in a 80286-based PC over a dirty phone line[1]; it like expecting an accomplished chef to prepare excellent soufflés with a rusty frying pan and a fork, or an artist to use a dime-store paint box. In these restricted circumstances, print has obvious advantages—it is widely available, relatively cheap, and the technologies involved in production and distribution, while often heavily computerized at points, typically still incorporate many mechanical elements.

Print has other features that make it valuable in settings where educational reform is central. Print is relatively permanent—it can be passed from hand to hand, viewed by many users without special equipment, it creates a record that may be tracked over time, and brings with it the weight of trustworthiness and authority that many electronic media still lack. In some cultural settings, such as Russia, print carries additional advantages: It is associated with an intellectual tradition, seen as the bearer and guarantor of culture, and may be seen by those in education as a kind of intellectual bulwark against chaotic forces of change thought to be working against the fundamental purposes of schooling and education.

The Role of Print in the Soviet Union

On one of Moscow's main streets, just a few yards from the Kremlin, there is statue of Ivan Fedorov, Russia's first printer. If one walks a few yards farther up the same street, one comes to the notorious Lubianka prison, the headquarters of the former KGB (state security police) and occasional home to many an arrested dissident writer under the Soviet regime. The coincidental juxtaposition of these two monuments is instructive, for it says something about the relationship between printed texts and the state in this distinctively Western-Eastern culture. For several hundred years, intellectual life in Russia was played as a cat-and-mouse game between those who wanted to write and publish their ideas and those who worried that the expression of those ideas in print would be dangerous to the state. Writers chafed, but in some ways also flourished, under the pressures that this put on them. A whole mode of printed discourse evolved—the so-called Aesopian language[2]—to allow an author to express critical ideas. Creative intellectuals were able to carry on a systematic and thoughtful dialogue on a whole range of contemporary Soviet social and political issues, while ostensibly discussing periods in ancient history, the problems of "decadent" countries of the West, or imagined societies on distant planets in the far future.

Tensions Between Printed Texts and Russian Politics

Historians have argued for years about the specificity of Russian culture and the unusual combination of elements that gave rise to the tensions around printed texts endemic in Russian society. This was a country that languished under the "Tartar yoke" of Asiatic despotism and domination at a time when most Western countries were dealing with the intellectual tensions of the Renaissance and Reformation. During the Enlightenment, when Western philosophers debated the rights of man and evolved new forms of representative government, Russians continued to live in a feudal society centrally controlled by an autocratic Tsar and the Orthodox Church. Basic literacy was always much less widespread in Russia than in the West; consequently basic cultural forms such as church rituals were founded on a tradition that was more oral and less written than in Western countries.[3]

Growing influence of central control. As Russian culture became more defined, the power of central authority grew with the efficiencies of eighteenth- and nineteenth-century technologies and so did the perceived power of printed materials. Tsarist censors were no less active than their future Soviet counterparts, and many writers in the nineteenth century suffered as they sought to work around the censor's requirements. In public life, individuals came to fear the power inherent in printed materials, and preferred to keep things "in the drawer" (out of print) rather than commit them to publication. If documents were dangerous, one would take care to have as few of them as possible. Soviet censorship mimicked many aspects of its Tsarist predecessor: It pressed writers to conform to standards set by central authority; it made printed materials scarce and important; and it established an official image of the society and its people.⁴

Under Stalin, this complex set of policies became codified as "Socialist Realism," a doctrine that specified how social themes and questions were to be dealt with in fiction. One variation on this position was known as the "Theory of Conflictlessness" (*Teoriia bezkonfliktnosti* in Russian) —the idea that in Soviet society, things were so perfect that there could be no conflict. How to write a novel in this vein, however, remained something of a puzzle, and many Stalin-era works were scoffed at by serious writers.⁵

Curious interactions of writing and censorship. Writers and the state censorship organs (*Glavlit* under the Soviet regime) existed in a curious kind of symbiosis. From the 1960s to the 1980s, a devoted corps of scribes typed and retyped editions of underground self-published (*samizdat*) materials, and then circulated the resulting smudgy carbon copies to friends and associates. Libraries that held any materials that might be considered dangerous or subversive—and, in the Soviet case, this included a surprisingly large number of works on educational psychology and pedagogy—kept these in a *spetskhran*, a special archive, to which access was by permission from appropriate authorities and misuse of which could lead to the loss of one's job or the chance to find a new position.⁶ Typewriters were said to have their individual distinctive "signatures" registered by the KGB, and access to photocopiers (the Latvians and Estonians provided the Soviet Union with relatively good models) was severely restricted. Print, in the Russian and Soviet tradition, was something dangerous, something to be controlled, something that had consequences.

State control in relation to education. Education in Russia was even more subject than other fields to state pressures that restricted and controlled the flow of printed materials. Communist ideology recognized from the start that education was a key link in the formation of the "New Soviet Man," and the task of assuring that only politically correct texts would be used in schools was consequently seen as an important one. Texts were developed centrally by the Academy of Pedagogical Sciences of the USSR, through its affiliated research institutes, and were thoroughly vetted by the Communist Party before being allowed into schools. Teachers' manuals went through the same process. As teachers made their way through pedagogical training, they were carefully counseled to avoid the path of unofficial approaches and unlicensed experimentation, and were encouraged to use the methods outlined in the approved manuals. Pedagogical spontaneity in the matter of text selection was not merely suspect, it was impossible; there were no alternatives available, no photocopiers in the schools, no paper to make the copies. Such publications as there were for teachers offered no alternative visions of what might be, only hortatory encouragement to do better what was already being done.

Educational experimentation in spite of central control. But these pressures on Russian educators did not dissuade the best of them from experimentation. Indeed, there was a long tradition among Russian teachers, extending back to prerevolutionary times, of creative dissent and protest against the excesses of the state (Seregny 1993). For many of these thoughtful, intellectual teachers, the control exerted by the state over the "means of educational production" represented just another challenge in what they saw as their true mission: to introduce their charges as completely and thoroughly as possible to the best works of Russian and world literature, science, and history. They did this out of an almost mystical conviction that this trove of material represented a kind of priceless intellectual and cultural capital, something that one needed if one were to become fully human and participate in the world cultural conversation. As many said, it was the basis for "universal human values." Printed materials, then, were not merely handy sources of educational content; they were rather critical elements in a process of cultural conservation, reproduction, and renewal that they saw as threatened.

Such was the cultural context in the USSR in the mid-1980s, at the beginning of the stormy period of social and economic change that came to be called *perestroika* (restructuring). Begun by Mikhail Gorbachev as an attempt to revive the moribund Soviet economy, perestroika soon affected society more broadly. One such impact was in education, perhaps the most static and (in the view of many progressives) moribund sector of the Soviet system.

Educational Reform Comes to Russia

By 1984, even Soviet government and Communist Party officials, long sanguine about the success of the educational system in providing both high-level scientists for the country's research institutes and efficient workers for the nation's defense plants, began to recognize that the existing structure was not doing a good enough job. Students in schools were often bored, at best, or actively alienated from the institutions of education and society at worst. Teachers, increasingly authoritarian in their approaches, often clashed with parents seeking to safeguard their children's psychological well-being.[7] Workers emerging from the system were trained in narrow specialties, lacked minimal initiative, and seemed unable to adapt to the increasing demands of technologically sophisticated industries. Education was organized hierarchically, with central ministries and committees dictating to lower republic-, regional-, and city-level organs; but the levers of command and control were so unidirectional that there was little feedback, little sense at the center of what was actually happening in individual regions and individual schools.

Teachers, selected mostly on the basis of ideological compliance and docility, had few avenues for professional growth or development. The networks of interconnected professional organizations (unions, subject-matter teaching associations, project-based efforts at local or national levels) and their media (newsletters, journals, annual meetings, etc.) were mostly lacking—a few state-controlled groups provided a facade of professional activity, but there was little of practical value to teachers, and few saw any need to become involved. Such information as there was flowed sporadically, via individual connections and links, from one small group of interested teachers to another. Those in higher administrative or research positions rarely expressed interest in the practical concerns of teachers or in thinking creatively about educational matters; those who did often faced official disapproval as "publicity seekers," "unwarranted critics," or, in extreme cases, "enemies of socialism." To say the least, this was not a professional milieu that encouraged its practitioners to think in diverse ways about their work.

Three forces combined to break this monopolistic, centralized, state-controlled and state-defined system of education in the mid-1980s: (1) the radical reorientation under a new, reform-minded editor of an existing newspaper for teachers, which provided them for the first time with a professional voice and glimpses of how the culture of teaching might be reimagined; (2) the chance, made possible by perestroika, to design and offer new kinds of professional development programs for educators, based on a set of powerful native psychological models; and (3) new models for powerful, theoretically sound curricular programs, primarily textbooks and related materials, realized through alternative programs of development, publication, and distribution. In all these cases, print-based materials were central. I will consider each briefly here, proceed to an examination of the persistence of print as a medium to encourage educational change, and then examine the particulars of the social and cultural context that made print central for Russian educational reform, with an eye to considering its role in other settings.

Teachers' Gazette as a Voice of Reform in the 1980s

In 1983, when Vladimir Fedorovich Matveev came to the position of editor-in-chief of *Uchitel'skaia gazeta* (*Teachers' Gazette*), the paper was uninspired "required reading" for administrators and bureaucrats in schools across the USSR. Many teachers read it as well, less because it offered useful advice or pithy insight on the practice of teaching than because it was simply all there was. With a thrice-weekly circulation of over two million copies, *Uchitel'skaia* made its way into all corners of the country. Matveev had been successful as an editor at a literary magazine for young children, and his charge from the Party and government bureaucrats who appointed him was to shake up the staid paper (although they could little anticipate how their mandate, issued before the real start of perestroika, would come to be interpreted by the fiery editor they put in charge). Matveev now had to figure out what to do with a publication that had little to do with the real professional lives of those it supposedly served.

In an effort to redefine the paper and aim it at the real problems of educators, Matveev hired a number of journalists; some, like Simon Soloveichik, had written for years about issues of education, but because they often differed with official Soviet orthodoxy, many were seen as mountebanks and charlatans by the educational establishment. Others, like Alexander Adamsky, were young teachers who were too enthusiastic and committed to fit well into the confined roles that Soviet schools and administrators had designed for them, and who wished to try their hands at journalism.

Innovation and the Teachers' Gazette. The paper, over the next five years, became something of a cause célèbre in the USSR, and not merely among teachers. Indeed, some intellectuals said that there were three "newspapers of perestroika" —true advocates of a more open and humane system: the general-interest papers *Moskovskie novosti* (*Moscow News*) and *Argumenty i fakty* (*Arguments and Facts*), and the educational paper *Uchitel'skaia gazeta*. From 1984 through the end of 1988, when Matveev was forcibly removed by hardliners in the Communist Party, the paper sponsored a remarkable series of events, contests, referenda on educational issues, and debates on a set of pressing national and educational questions. Journalists discovered and held up as popular examples a series of "innovators" (*novatory* in Russian), teachers who worked not according to the standardized recipes imparted by the state pedagogical institutes, but rather according to their own convictions and ideas of what it meant to be both educated and human. A national educational conference organized by the government and intended to foster support for the ailing Soviet regime through education was postponed, new delegates selected and a new, wide-ranging agenda established based on a campaign initiated by the paper.

Thoughtful teachers and concerned researchers, brought together on the paper's initiative, drafted a manifesto, the "Pedagogy of Cooperation," which suggested that teachers could teach and students could learn in an atmosphere where coercion and threats were not parts of the everyday environment. Later, these same ideas served as the foundation for reform efforts under the new educational administration of the Russian Federation after the breakup of the USSR in 1991. The developments of this period have been characterized by many who were involved as the start of a "social-educational movement" (*obshchestvenno-pedagogicheskoe dvizhenie*), and many educators still refer to their participation or membership in "the movement."[8]

The rise of teacher professionalism. Throughout this remarkable period, *Uchitel'skaia* published not only position papers and accounts of the work of innovative teachers, it also encouraged teachers to see their work as a something more than a job or mere support for the socialist state. The paper, under Matveev's urging and direction, gradually articulated and encouraged teachers to adopt a new image of teaching as serious and professional, but above all as a fundamentally ethical human activity. Teachers responded to this call with a remarkable thirst, and when the paper sponsored a meeting of teachers in Moscow in October 1986, the level of interest was strong enough that Matveev charged Alexander Adamsky to see if similar interest existed in other parts of the country. (Matveev, Soloveichik, and Adamsky had debated what level of interest would be minimally needed to launch the effort; they decided that they would need to see at least thirty Moscow teachers at the first meeting. More than 400 showed up.) Within two years, there were nearly 500 teachers' clubs, many under the name "Eureka" (*Evrika* in Russian), operating in various cities and regions of the Soviet Union. Those active in the movement began to refer to it as "The Union of Teachers," after a real union that existed in prerevolutionary Russia, and in pointed distinction to the official State Union of Educational Workers.

Matveev, Adamsky, and Soloveichik went to discuss the new movement with the head of the State Committee on Public Education, Gennady Iagodin, and requested permission for the group to operate legally and publicly as an alternative pedagogical establishment. Iagodin, a competent late-Soviet-era bureaucrat, who was worried about establishing precedent in the still centralized Soviet school system, eventually gave official permission for the group to continue its activities, but only on condition that the word "Creative" be added to the title. The Creative Union of Teachers was thus born, a new structure for Russian educators and the first independent educational structure in the USSR focused on professional growth and development.

Drowning the Voices of Reform at the Close of the 1980s

The discussions and festivals organized by the Creative Union lasted only a brief time—from 1987 through 1990. Internal pressures, lack of managerial expertise, and the difficulties of communicating with a diverse membership spread across the territory of the USSR made it impossible to support the organization's fledgling structure, regardless of the levels of enthusiasm among its members. At a meeting in the autumn of 1991, the Union essentially self-destructed, and although remnants of it have continued to the present, the founding figures and many thoughtful teachers left at that point. In its brief lifetime, the Union, organized around a powerful set of new ideas, became a rallying point not only for teachers, but also for progressive intellectuals throughout society. Discussion of educational issues found a place on the pages of serious intellectual journals and papers; television shows featuring members of the intelligentsia debating pedagogical questions became immensely popular; the books of Simon Soloveichik, printed in large editions, sold out and became "bibliographical rarities" (see, e.g., Soloveichik 1989a, 1989b); many spoke of the coming "pedagogization of society," initiated by Matveev and his co-workers at *Uchitel'skaia gazeta*.

But the Communist Party was still strong, and its leaders could not brook the kinds of criticism that Matveev increasingly leveled at them from the columns of his paper. At the end of 1988, the paper was officially made an organ of the Communist Party, and Matveev was removed as editor. In 1989, under the new editor, Communist-Party-appointed Gennady Seleznev, coverage reverted to bland descriptions of official pronouncements and reportage on noncontroversial topics. While the paper improved under its next editor, Petr Polozhevets, it never quite regained its old spark.[9] In the stressed economic conditions of post-Soviet Russia, press runs for *Uchitel'skaia* have fallen from over two million to just over two hundred thousand, and its frequency from thrice-weekly to just once per week (although it has added more pages).

It was not until 1992, when Soloveichik was able to found *Pervoe sentiabria* (*September First*), a new educational newspaper for teachers, that there was again an independent and strong voice for educational change in Russia. Soloveichik, who always made respect for teachers the hallmark of his writing, devised this as a motto for the new paper: "You're a brilliant teacher; you have wonderful pupils!" At first, the paper also carried on its masthead a memorial to Matveev's *Uchitel'skaia*: "The original values, the original traditions." *Pervoe sentiabria*, which is published three times per week, is also supported by fifteen supplements on all the subjects in the academic curriculum, plus such topics as children's health and school administration. Its combined press run is about the same as that of *Uchitel'skaia gazeta*.

Programs for Teacher Reorientation

How could the kinds of ideas suggested by Matveev's paper and the groups it spawned be carried into practice? Clearly, there was a need to launch programs to reorient teachers to the new ways of working and thinking that were being proposed and discussed in the print media. And, this being Russia, it was centrally important that the programs represent a sound theoretical basis—Russian teachers expected that, and they had a rich tradition of work on pedagogy and educational psychology on which to draw (even though, for most of them, their experience with that tradition had been second hand, as interpreted through official Communist Party education ideologues).

Struggle for Influence within Academic Circles

In the 1980s, a contest that had been brewing for years inside Soviet academic psychology came to a head in the dispute over who should lead the Psychological Institute of the Academy of Pedagogical Sciences of the USSR. Vasily Davydov, a progressive educator and one of the few researchers whose ideas were known and respected by working teachers, had been tolerated as head of that institute for a number of years, but was ousted because of his interest in and support of the ideas of the Russian psychologists Vygotsky, Luria, and El'konin (and perhaps also because he ran the Institute with fearless disregard for the anti-Semitic preferences of many of those elsewhere in the educational establishment). As perestroika accelerated, Davydov's views once again became acceptable to some (if not all) in positions of authority, and he resumed that position. This made it possible once again to bring the ideas of Vygotsky and the others to the fore in programs designed to prepare teachers to work in new ways.

Vygotsky's "cultural-historical" approach to psychology and pedagogy was attractive to many progressive educators because it stressed the social origins of development and knowledge, and because it emphasized the value of "reflection" as a way of generating personally valuable new insights that could be translated into action.[10] Vygotsky's ideas and those of his followers and other renowned psychologists such as G. P. Shchedrovitsky came to serve as one basis for new programs to prepare teachers offered by a variety of groups—Adamsky's "Eureka Open University," Petr Shchedrovitsky's "School of Cultural Policy," Yuri Gromyko's Moscow Academy for the Development of Education (MARO), and the International Psychological College of the Psychological Institute (headed by Vitaly Rubtsov).

The specific views of education espoused by these groups vary, but they share some common elements. There is a focus on the teacher's role as participant in and "collaborative developer" of knowledge with the pupil ("developmental instruction" is the phrase often used as a description of this approach). With this new emphasis, the old Soviet didactic and highly teacher-centric approach recedes into the background. The view of the curriculum itself changes, with teacher and student taking more responsibility for defining the topics that will be studied and the approaches that will be taken in working on them. Models for the organization of the school and its relationships with parents and students also change, with more responsibility typically allotted to parents and to new school councils. As more and more schools have moved toward this approach to teaching, a new Association of Developmental Instruction has arisen to provide coordination and guidance. The fact that some 2,500 of the country's schools now belong to the association suggests that it may become the theme around which educational reform in Russia will grow.

The Role of Print in the Rise of Activity Theory

The print tradition of educational psychology provided inspiration for the Russian educational reformers of the 1980s. As restrictions on publication were relaxed, the original works of educators and psychologists such as Vygotsky, Shchedrovitsky, Luria, and Sukhomlinsky began to be republished, creating no less a stir in Russian educational circles than did the republication of Alexander

Solzhenitsyn's novels among intellectuals generally. The reformers also drew on interpretations of the pedagogical and psychological tradition that depended on a unique view of how learning takes place, its powerful association with activity, the interaction of one human being with another, and the basis for that interaction in historical and cultural traditions and the systems of signs and symbols that those traditions left in their wakes. A whole technology for carrying this image into practice arose in the late 1970s and 1980s. Known as "organizational-activity games" (*organizatsionno-deiatel'nostnye igry* in Russian), these large-scale simulations were run in a number of major Russian cities and for various purposes—local educational development, regional social and economic planning, improvement of working patterns and relationships within organizations. These games almost always had some educational component, inasmuch as their originators were psychologists who saw education (in a broad sense) as essential to their work. Participants typically came together for several days or a week of sixteen-hour sessions, and expected to have their ideas powerfully challenged. They were not disappointed, as the usual method of challenge, attempted solution, confrontation, and critique left most participants exhausted after a few days. The process of "reconstructing" people's image of the world and their professional relationships was not all that different from the approaches used in basic military training, prisons, or initial medical or legal training.

These methods are used extensively today in the teacher training sessions run by the various groups noted above. There are demonstrations and models of new teaching approaches, opportunities to discuss and challenge those new models, and guided efforts by the organizers to encourage the participants to challenge their taken-for-granted assumptions about the shape of the pedagogical world. These challenges ("reflection" sessions, in Vygotskian parlance) are often extremely stressful for those involved, leading to angry outbursts, heated confrontations, tearful breakdowns, confused withdrawals, or sudden, unexplained departures.

A number of specific techniques, some themselves involving further use of printed materials, are used to move the participants toward reconsideration of long-cherished "truths" of the pedagogical method. Adamsky's Eureka group, for example, employs a graphic artist who sits in most sessions and draws cartoon representations of the participants, sometimes in especially revealing or difficult moments. These drawings, together with appropriate captions from the subject's own remarks, and accompanied by other written comments about the session prepared by fellow teacher-participants as well as by the organizers, are posted for all to see. Sometimes by the end of a meeting the entire public area of the hotel or conference center is covered with drawings, replies, reposts, and commentary. It is a recapitulation of the old Soviet phenomenon of the "wall newspaper," a semi-spontaneous device for propaganda delivery and local mobilization.[11]

An Unsettled Future for Russian Educational Reform

It is difficult to tell at this point what the overall impact of these new programs for teacher retraining will be. While the absolute number of teachers from around Russia and the former Soviet Union who have participated is relatively large, their relative impact on the country's sprawling educational system is still quite small. Indeed, in the early years of the movement, many teachers who participated in the various seminars and colloquia run by the reformers did so on their own time and spent their own money. While this resulted in a considerable dispersal of the group's ideas, it also weakened their impact at the local level, since it was unlikely that more than one or two teachers from any given locality would have been exposed to the new ideas. Later, Adamsky and some of the others began to require that seminar participants come in teams from individual schools or regions, but this step also coincided with the implosion of local educational budgets which drastically curtailed travel for most educators.

The new programs for teacher retraining and development count heavily on the historical record of Vygotsky and others as retrieved from the special archives of Soviet-era libraries and newly published in large press runs, and they are instantiated in the programs and books that the various groups noted above have created. These programs also capitalize on new interpretations of culturally rooted uses of printed materials such as the wall newspaper. But the powerful impact of these new programs of teacher reorientation on those who participate in them cannot be understood without noting the connection of the new programs with a historically honored and previously persecuted print tradition of educational psychology native to the country. This tradition allowed the reformers to claim legitimacy for their work.

Need for New Textbooks in the New Era

Getting teachers to work in new ways is one thing; giving students the tools to do so is something quite different, especially in a country where the entire textbook industry was centrally controlled and hardly a model of efficiency. In the Soviet era, all texts emerged from internal development efforts in the Academy of Pedagogical Sciences. Texts, once approved, were published in large press runs and distributed (inefficiently) throughout the country. Titles frequently remained in use for years, and there were heavy ideological "loadings" not only in such obvious subjects as history and literature, but also in the arts and sciences.

Efforts to create new texts in the 1980s and 1990s ran directly into the heritage of this Soviet system: lack of alternative channels for the solicitation of manuscripts, lack of a structure for independent review of the same, lack of capacity in printing plants (to say nothing of endemic paper shortages and press breakdowns), and an almost completely ineffective distribution system. In the late 1980s, several editions of history texts were created as it became allowable to address a wider range of subjects; most never made it into the schools, and teachers were at one point told to skip final exams in history and instead to conduct class discussions based on current issues as illustrated from newspaper and magazine clippings.

Curriculum reform. In order for new texts to develop, there first had to be a new conception of the curriculum that would permit alternative views of curricular content to emerge and that would foster the creation of divergent materials. The groundwork here was laid by VNIK-Shkola (*Vremennyi nauchno-issledovatel'skii kollektiv Shkola* or Ad Hoc Committee on the School), a group that was created on the initiative of Gorbachev's own science advisor, Evgenyi Velikhov. While VNIK formally was administered under the auspices of the USSR Academy of Pedagogical Sciences, and thus had at least minimal legitimacy in the eyes of the existing power structure of the educational bureaucracy, its real power (and thus its authority to urge change) came from Velikhov's links to the USSR Academy of Sciences (many of whose members served as collaborators on VNIK's various working groups and commissions), and ultimately from Gorbachev himself. As such, it was a good example of a reform-oriented group which was created outside the structure it was intended to reform, which had powerful external sources of support, and which ultimately extended its charge beyond what was originally contemplated.

Heading VNIK's effort was a historian of Russian education, Eduard Dneprov, whose previous work had focused on the political and social role played by radical teachers during the immediate prerevolutionary period. Dneprov was an ardent "Child of the Twentieth Party Congress" (the famous meeting in 1956 where Stalin was semi-publicly denounced for his excesses), and worked feverishly to involve not only educators but also scientists and public intellectuals in a wide-ranging discussion about the curriculum, the purposes of schooling, and the ways in which new approaches to education should be developed. The work of VNIK, as put forward in several dozen reports and pamphlets covering all areas of the school curriculum during 1988 and 1989, provided a clear image for the reformers of what was needed, what had to be done, and to a somewhat lesser extent, how to go about achieving those ends.[12]

Lack of infrastructure and capital. Merely having an opening for the preparation of new school materials, however, was no guarantee that these would actually appear. Even in the relatively open environment created by Gorbachev in the last days of the Soviet Union, there was much that remained difficult or impossible. There was no infrastructure capable of soliciting manuscripts, editing and preparing the best of these for publication, and printing and distributing these to schools. In the late USSR, there were shortages of everything, from the paper for printing the books, to places in press publication cues, space on strained transportation systems to carry books to local regions and schools, channels through which to notify teachers that new texts were ready, and funds to provide for the retraining of teachers to use the new materials. In this sort of chaotic environment, there was no point in trying to depend on the existing structures, as everyone knew that they were completely incapable of providing what was needed.

Emergence of alternative publication outlets. As the existing publishing houses for instructional texts proved themselves unable to adapt to new conditions, several groups emerged to fill the need. Two bear special mention here: MIROS (the Moscow Institute for the Development of Educational Systems), and *Novaia Shkola* (New School). MIROS, under the direction of Alexander Abramov, produced several hundred new titles between 1994 and 1998, and carried out the process from soliciting manuscripts through production and distribution. Some of the titles, especially on subjects such as ancient history and art, have been so well received that they have become popular best sellers as well as being used in schools. MIROS has benefited from money provided by George Soros's Open Society Institute, but has also learned that books provided for nothing are not valued by the recipients as much as those for which one is charged, and so the firm now requires schools to pay for the texts.

Another alternative publishing organization, New School, directed by Vladimir Girshovich, is somewhat more pragmatic and market oriented than MIROS, and has created a whole series of practically oriented pedagogical guides and texts for schools. In addition to these, the firm also offers seminars and publishes a newspaper, *Pedagogical Kaleidoscope*, and a teacher-oriented magazine, *World of Education*. The content of all these publications is interesting, for it seems calculated to appeal to younger teachers, who might find the ideas of developmental instruction and Vygotskian theory too academic and rooted in the past. Nonetheless, in the current situation, all these operations for preparation and dissemination of texts depend on the use of printed materials.

Still other channels are used to prepare and disseminate new texts. The various educational newspapers (*Uchitel'skaia gazeta*, but also such publications as the conservative *Pedagogicheskii vestnik* and the various supplements of *Pervoe sentiabria*) all publish lesson plans for teachers and, occasionally, entire model texts in multiple installments. There are also a number of regional educational development centers that have taken it on themselves to prepare series of texts in particular disciplines (history and local culture are especially popular, but there are locally developed texts in such fields as religious studies, natural sciences, and mathematics as well). The center in Nizhnyi Novgorod (formerly Gorky) has been especially active, and its series on citizenship education (*grazhdanovedenie*) has been seen by many in Russia as a model for how local development of educational materials can proceed. As with many other educational development efforts in the former USSR, the resources made available by George Soros's Open Society Institute have been instrumental in much of this work. Nonetheless, increasing textbook availability continues to be a central theme for the ministry in its attempts to bolster the weakened educational system (Kinelev 1997).

Interestingly, this search for new texts sometimes extends into the past. Several publishers have reprinted prerevolutionary Russian textbooks and found a ready market for them among the new *lycees* and *gymnasia* that have sprung up among Russia's former elite secondary schools. Why teachers would willingly reach back more than eighty years for models with which to inform current pedagogy is something of a mystery to an outside observer, but it apparently makes perfect sense to some Russians. The lure of a long-unavailable prerevolutionary history text is strong; the reasoning among some educators may be that since they were forbidden to read these books for so long, there must have been something very powerful and worthwhile in them. The notion that there might be a more useful or more appropriate contemporary interpretation of Russian (or world) history seems less than compelling for these educators.

As in American and other Western schools, textbooks are important for a number of reasons—they are handy for students and teachers to use; they offer a synopsis of a field that is explicitly designed to be accessible to novices; and they can be relatively easily redesigned and republished as a field changes. In Russia, the need for new texts was compounded by the heritage of ideological influence over existing materials, and the consequent need to create new models of content and new approaches to the teaching of such central subjects as history, economics, and literature. The design and creation of new textbooks in the early 1990s, supported by a radically new vision of a traditional liberal curriculum, was a powerful moving force for Russian educators to reform their system and their schools.

The Fate of Print in an Electronic World

It is clear that printed materials remain an essential feature of the Russian educational landscape, and will continue so for many years to come. The best efforts by native and Western entrepreneurs will not be able to bring Internet connections and ISDN lines to any meaningful percentage of Russian schools in the near future, and even if they did, economic conditions are not likely to improve sufficiently to allow schools to purchase the hardware needed. Print is therefore a pragmatic, economic necessity.

But printed materials clearly have another, perhaps deeper set of connotations in the Russian context. They have historically been at once the repository of important cultural ideas, the source of radical (hence potentially dangerous) new information, and the channel through which serious intellectuals communicated with their peers and audiences (audiences that historically, especially in prerevolutionary Russia, included teachers to a greater degree than was the case in the United States). Print was the vehicle for ideas of power, a medium that invited careful attention from both the government censor and the concerned citizen. It was the incorruptible spring from which centrally important theories could be drawn (or redrawn) to replenish the parched body politic. All of the printed materials used in education partook in this same tradition, from the textbook used to teach Russian language in the first grade, through the physics text for university students; from the manuals for teachers that were designed and circulated to assure that there was minimal variation among classrooms, through the traditional academic journals; and from the rural teacher's paper that appeared a few times per year to the thrice-weekly official All-Union education newspapers.

The results of this pattern of cultural specifics are clear to any critical observer who visits Russia, and Russian schools in particular. There is a near reverence for printed materials, especially those that are seen as bearers of the intellectual and cultural tradition. Texts are preserved and treated nearly as objects for reverence, not resold to bookstores as soon as the academic year is over. In Russian homes, works of literature are sought out and carefully added to the large, glass-fronted bookcases that are the feature of nearly every Russian living room. Journals and newspapers are carefully folded flat and preserved in bundles and stacks, tucked away on the top shelves of closets and behind the skis in the hallway. Documents, the country's heritage of intellectual wealth and political terror, are saved and treated as the bearers not only of information, but also of a kind of strange and impenetrable mystical authority.

The present day has brought challenges to this tradition. Publications available from street kiosks in the new Russia are as likely to be detective thrillers, conspiracy-mongering anti-Semitic diatribes, or how-to books on computers or marketing, as the traditional classics of Russian and world literature so beloved in the past. Newsstands are as likely to carry lurid sex manuals or Russian editions of *Playboy* as they are the traditional newspapers that cater to teachers and intellectuals. Computer communication links, while still unusual, are increasingly available to urban Russians.

Whether this traditional Russian pattern of reverence for the printed word can sustain itself over the long run in a context where Western influences and the penetration of new electronic forms of communication are making their presence felt more strongly every day is an intriguing question. The Russian past, with its unusual combination of forced censorship of the written word and the power attributed to it by intellectuals, may reassert itself to reshape and provide new weight for electronic means of communication, or perhaps the Western model of "disposable text" will prevail. These are questions yet to be answered. Watching what will happen may allow us to test whether cultural forces and traditional assumptions can hold sway in the face of what in the West are typically seen as "inevitable" market forces.

Notes

1. A school that has access to financial resources and is located in a medium or large city might have up-to-date machines. In rural areas (and many of Russia's schools are rural) one would be lucky to find a working computer more advanced than an Apple II clone or a 286 PC. And the problem of "dirty" phone lines is a major block to reliable modem communication.

2. This is a reference to what *Webster's Collegiate Dictionary* defines as language which conveys "an innocent meaning to an outsider but a hidden meaning to a member of a conspiracy or underground movement."

3. Billington (1966) provides a good summary of the trends at work in Russian culture; see Ong (1982) and Jaynes (1976) on characteristics of oral cultures.

4. On the philosophical origins of Soviet propaganda and censorship, see DeGeorge (1966); on its practical application, see Remington (1988); on the distinctive Russian passion for books, see Mehnert (1983).

5. See Gorokhoff (1959) and Gromova (1995) for details on the development and practice of Soviet censorship; Tertz (Andrei Siniavsky) 1960 offered a useful critique of socialist realism.

6. See, for example, Vasily Davydov's comments on the KGB's administrative control over L. S. Vygotsky's 1926 *Pedagogical psychology* in Moscow's Lenin Library (Davydov 1995.)

7. Surveys done in the late 1980s showed that 37% of teachers reported "sharp conflicts with individual students," while nearly 60% noted the chief barrier to educational reform as "alienation of families from their children's education." *Shkola-1988*: 8, 19.

8. On the origins and work of "the movement" see, for example, Dneprov, Kasprzhak, and Pinsky (1997); Gazman and Kostenchuk (1995); Tubel'sky (1989); and *V poiskakh* (1993).

9. Seleznev, interestingly, went on to become the last Soviet-era editor of the Communist Party newspaper *Pravda*, then, after the collapse of the USSR, became speaker of the Russian Duma, or parliament.

10. For treatments of Vygotsky in English, see Kozulin (1990), Moll (1990), Newman and Holzman (1993), and Wertsch (1985).

11. See Kenez (1985) 237-240 for a description of the origins of wall newspapers and their uses as propaganda devices.

12. For a comprehensive account of the work of VNIK, see Eklof and Dneprov (1993).

References

Billington, J. H. 1966. *The icon and the axe: An interpretive history of Russian culture*. New York: Knopf.

Davydov, V. V. 1995. The influence of L. S. Vygotsky on education theory, research, and practice. *Educational Researcher*, 24(3):12-21.

DeGeorge, Richard T. 1966. *Patterns of Soviet thought*. Ann Arbor, MI: University of Michigan Press.

Dneprov, E., Kasprzhak, A., and Pinsky, A. Eds. 1997. *Innovatsionnoe dvizhenie v rossiiskom shkol'nom obrazovanii* (The innovative movement in Russian school education). Moscow: Parsifal.

Eklof, B., and Dneprov, E. 1993. *Democracy in the Russian school*. Boulder, CO: Westview Press.

Gazman, O. S., and Kostenchuk, I. A. Eds. 1995. *Gumanizatsiia vospi taniia v sovremennykh usloviakh*. (Humanization of upbringing under current condition). Moscow: Russian Academy of Education, Institute for Pedagogical Innovations.

Gorokhoff, B. I. 1959. *Publishing in the USSR*. Slavic and East European Series, vol. 19. Washington, DC: Indiana University Publications/Library Resources.

Gromova, T. V. 1995. *Tsenzura v tsarskoi Rossii i Sovetskom Soiuze: materialy konferentsii 24-27 maia 1993* (Censorship in tsarist Russia and the Soviet Union: Materials from the Conference of May 24-27, 1993). Moscow: Rudomino.

Jaynes, Julian. 1976. *The origin of consciousness in the breakdown of the bicameral mind*. Boston: Houghton Mifflin.

Kenez, Peter. 1985. *The birth of the propaganda state: Soviet methods of mass mobilization, 1917-1929*. New York: Cambridge University Press.

Kinelev, V. G. 1997. *Doklad ministra obshchego i professional'nogo obrazovaniia Rossiiskoi Federatsii V. G. Kineleva "Ob itogakh raboty Minobrazovaniia Rossii v 1996 godu i osnovnykh napravleniiakh deiatel'nosti na 1997 god"* (Report of V.G. Kinelev, Minister of General and Professional Education of the Russian Federation, "On the results of the work of the Russian Ministry of Education for 1996 and basic directions for activity in 1997"). *Vestnik*

obrazovaniia [Education Herald]. 3: 78-110.
Kozulin, Alex. 1990. *Vygotsky's psychology: A biography of ideas.* Cambridge, MA: Harvard University Press.
Mehnert, Klaus. 1983. *The Russians and their favorite books.* Stanford, CA: Hoover Institution Press.
Moll, Luis. 1990. *Vygotsky and education: Instructional implications and applications of sociohistorical analysis.* New York: Cambridge.
Newman, Fred and Holzman, Lois. 1993. *Lev Vygotsky: Revolutionary scientist.* New York: Routledge.
Ong, Walter. 1982. *Orality and literacy: The technologizing of the word.* New York: Methuen.
Remington, Thomas F. 1988. *The truth of authority: Ideology and communication in the Soviet Union.* Pittsburgh, PA: University of Pittsburgh Press.
Seregny, Scott. 1993. Teachers, politics, and the peasant community in Russia, 1895-1918. In *School and society in tsarist and Soviet Russia* edited by Ben Eklof. New York: St. Martin's Press.
Shkola-1988. *Problemy, protivorechiia, perspektivy* (School-1988. Problems, contradictions, prospects). Moscow: VNIK-Shkola (State Committee on Public Education).
Soloveichik, S. L. 1989a. *Pedagogika dlia vsekh* (Pedagogy for everyone). Moscow: Detskaia literatura (Children's Literature).
———. 1989b. *Vospitanie po Ivanovu* (Upbringing Ivanov-style). Moscow: Pedagogika (Pedagogy).
Tertz, A. (Siniavsky, Andrei). 1960. *On socialist realism.* New York: Pantheon.
Tubel'sky, A. N. 1989. *Shkola samoopredeleniia—pervyi shag* (The school of self-determination—The first step). Moscow: VNIK-Shkola, USSR Academy of Pedagogical Sciences.
V poiskakh novogo soderzhaniia obrazovaniia (In search of a new curriculum). 1993. Krasnoiarsk: Krasnoiarsk State University.
Wertsch, J. 1985. *Vygotsky and the social formation of mind.* Cambridge, MA: Harvard University Press.

Chapter Five

Women, Telephones, and Subtle Solidarity: A Counternarrative

Sousan Arafeh

Much scholarly research indicates that solidarities[1] are often forged by women[2] within the quasi-public communicational and relational space afforded by telephone technology. However, this research tends to characterize the telephone as a private and gendered technology of containment through which women's physical, face-to-face, and political interaction is seriously and severely limited. I argue that although aspects of this ring true, the telephone is a place where solidarities of many kinds are both formed and broken. I term these solidarities "subtle and supple." They are subtle because they are difficult to perceive. They are supple because the forms they take are complex and highly situational.

This chapter critically reviews liberal feminist, social scientific, and psychological literature on women's telephone use. What becomes immediately evident is that this scholarship emerges from and perpetuates the dominant "containment" discourse of women and phones. However, theoretical developments regarding notions of women, spaces of social and political interaction, and the telephone as a technology call into question characterizations of telephone containment. To begin to identify and characterize some of these developments I explore scholarship on phone use by different nondominant populations such as gay men, sex workers, psychics, or other forms of work that adopts a postmodern theoretical perspective. These kinds of approaches to telephone technology and use form a counternarrative to the dominant critical literature on women and phones. The counternarrative I offer suggests that telephone identities and practices are much more fluid and complex than is typically portrayed, and that the space of the telephone is more public and more political than previously construed.

The discourse of women and telephones to date continues to draw heavily upon the traditional containment perspective which paints a fairly bleak picture of the phone as a tool and space for political action and affiliation. Thus, further study about women's and other nondominant communities' communicational and relational phone practices is warranted. Not only should such scholarship prove useful for better understanding the discourses and practices of how nondominant populations use and have used telephone and other communications technologies; it should also help to better understand emerging discourses and practices relative to new communications technologies such as the Internet.

Telephones, Communicative Space, and Gender

In the United States, the telephone is typically conceived of as a personal communications appliance which can be found in most households, businesses, and public spaces. The telephone has value because it is connected with other telephones—and thus people—around the world by national and global telecommunications networks. Long a staple of U.S. society, telephone hardware and service are affordable and accessible for most people, especially since federal regulations require universal service. Currently, for example, telephone penetration is 94% in U.S. households and 96% in Canada. This is not to say that there aren't significant disparities in who has access to in-home phones. In November 1997, telephone penetration reached 77.2% for households with annual incomes below $5,000, and although whites by far dominate the overall household penetration rate at 95%, both black and Hispanic household telephone subscribership weigh in at 86%. Even unemployed adults have significant access to in-home telephones (86.8%); only 8.6% less than employed adults (95.4%) (Find/SVP 1997).

Visual Representations

The telephone is depicted and often experienced as a transparent, almost magical, way to "reach out and touch someone." Visual representations of telephones and telephone users typically show shot/reverse shot scenarios of individual people talking with other individual people, over short and long distances, within a highly private communicational space. And yet, identity is a bit more fluid than these depictions would suggest, since voice is the point of reference and a person's visually perceived, somatic presence is displaced. The privacy and ease of use afforded by the telephone are regulated and practically supported by telephony's common carrier status, which protects users from wiretapping and other surveillance, except in the most severe circumstances, and by its technical ability to transmit voice in real time with minimal modification. Phones are easy to operate, do not require the degree of computational skill or English language proficiency that the Internet does, and have highly reliable connections. New developments in cellular and wireless communications afford increased physical and geographic mobility.

These kinds of point-to-point representations of the telephone are misleading, however. While it is true that longstanding regulatory measures have ensured that the telephone's broadcast[3] capability is not as robust as that of the Internet, for example, it is also true that these two technologies depend upon the same networks of technical infrastructure. Telephone systems are fundamentally networks that can be—and are—used to forge and span communities. And yet, telephones are rarely represented in this way. This would suggest that telephones are inefficient for communicating broadly among many people. However, as anyone who has passed on news of interest in a community by phone or has participated in a "phone tree" knows, the telephone is a powerful communicational tool with the potential of rapid and significant reach. It seems to be the case, then, that dominant depictions of telephones and telephone use offer only very limited information about what phones are and how people use them. These dominant depictions are highly constitutive of the discourse of telephones and phone use—one that needs to be problematized, as I noted above.

Gendered Technology

In keeping with research that technologies become "gendered" through patterns of socialization and use (Morley 1986; Spender 1995; Wajcman 1991), it is possible to see how dominant narratives and representations of the domestic telephone characterize its spaces, uses, and users in gendered terms. The phone is intimate. It connects the private sphere of the home within the private space of the phone connection—the private line. The phone is also immediate and emotional—perfect for frequent and involved conversations and social connections which typically occur in the private, personal space of the home. Although phone space and use is gendered and portrayed as an extension of (often women's) private communicational space, should one take this depiction at face value? What happens if a person chooses not to honor the characterization of the telephone as bounding and spanning a private sphere and, instead, construes this space as something more public?

Public and Private

Theoretically, the distinction between public and private spaces, or spheres, is well established in Western thought. Important for the purposes of study are the ways in which these distinguished spheres are attributed political power and efficacy. Public sphere theory as propounded by scholars such as Kant[4] and Habermas[5] projects and maintains the Cartesian mind/body distinction onto social and political space. It constructs the public sphere as a rational space of communion and communication, the sphere of business and culture. Important, instrumental exchanges of information and goods occur within it, many of which are necessary for successful political alliances and actions. In contrast, the private sphere is the sphere of nature, emotion, and nurturance. In and through it, the day-to-day relational activity of physical and emotional growth, support, and sustenance takes place. Activity within this realm, although possibly constituting threshold activity which substantially supports formal political action, is construed as not itself formally political. Basic characterizations and assumptions of public-sphere theory as it addresses social and political activity are not useful for the purposes of this chapter, since I want to argue that the private space of the point-to-point phone line is a highly political space where alliances and actions are conceived and executed daily. However, drawing upon Fraser's (1992) modifications to public sphere theory, it is possible to claim that telephone networks comprise a social and relational space that is both quasi-public and highly political.

Fraser suggests that public-sphere theory is theoretically unable to support formal, democratic, political action because it is based on the exclusion of certain people and populations from the social sphere where political action supposedly takes place. The primary strategies employed to manifest these exclusions include "bracketing social inequalities" (i.e., rhetorically proceeding as if there are no distinctions among people), and making their access to means of participation structurally and interpersonally difficult. Such strategies of exclusion should draw attention to those populations being excluded or perceived as needing to be excluded. Fraser attempts to account for these populations' experiences by positing that democratic activity also takes place among multiple publics, including "subaltern counterpublics"—groups that exist outside the formal public sphere and which, when foregrounded, decenter "bourgeois" cultural dominance in the dominant public sphere. Subaltern counterpublics are "parallel discursive arenas where members of subordinated groups invent and circulate counterdiscourses to formulate oppositional interpretations of their identities, interests, and needs"(Fraser 1992, 123).

To illustrate her concept, Fraser holds up "the late 20th century US feminist counterpublic" which, by using a "variegated array of journals, bookstores, publishing companies, film and video distribution networks, lecture series, research centers, academic programs, conferences, conventions, festivals, and local meeting places" were able to identify and name experiences of "social reality" particular to its members (e.g., oppressions). The process of communicating in alternate public spheres and the process of inventing a new language for their experiences allowed these women to recast "needs and identities, thereby reducing, although not eliminating, the extent of disadvantage in official public spheres" (123).

What Fraser's modification to public-sphere theory allows is the possibility of different spheres of activity, and the possibility of important political activity within these alternate spheres. In this chapter, I construe what is traditionally thought to be the private space of the telephone line/network as a "quasi-public" space wherein communicational, relational, and thus political activity takes place in ways that are so subtle and supple that most attempts to analyze it have only scratched the surface, particularly as these spaces and activities pertain to women and other nondominant populations. The review of the literature that follows, then, is one attempt to understand ways to explore the complexity of who and what is interacting in the technological, communicational, quasi-public space of the telephone.

Historical Relation of Women and Telephones

Since the telephone offers a quasi-public sphere of communication that is gendered female, it will be useful to consider the historical location of women in relation to it. There is not a great deal of historical research on women and telephones. As those who undertake work in this field note, this is a surprising fact considering the degree to which women and phones have historically been associated— as switchboard operators, secretaries, gossips, or relationship hopefuls waiting impatiently by the phone (de Sola Pool 1977; Fisher 1992; Martin 1991; Moyal 1992; Rakow 1992; Sarch 1993). The history of women and phones has primarily two components: the history of women as phone workers, which is rich and well-documented from a variety of perspectives,[6] and the history of women as phone users —one surprisingly limited in documentation. While I do not want to underplay the political, economic, and cultural effects that women in the United States and overseas have had as telephone operators and hardware and component manufacturers, or the exploitative nature of these relationships, the aspect of use concerns me in this chapter.

Quite early in the telephone's history, a discourse of women's "natural" abilities and affinities with the phone emerged—one that helped gender the technology and likely affected both male and female perceptions and uses of the phone. The discursive effect of phone company campaigns to recruit female operators may also have contributed to the image that women and phones were somehow naturally connected. These campaigns circulated images of courteous young women to boost the respectability of the phone itself and the pleasantness of the phone experience (Maddox 1977; Martin 1991; Rakow 1992; Langer 1972). Consider this quote from an early 1900s brochure developed by the New York Telephone Company in the 1910s:

> Among all who attend to our general needs, few are more important than the telephone operator. She establishes and controls the channels of the nation's speech. Presidents, senators, bankers and captains of industry lift their telephone receivers and ask for her services. Yet she also serves with equal efficiency the smallest merchant or the most humble citizen. With the aid of the wonderful organization of wires and workers, of which she is a part, she contributes her share toward the advancement of the nation and makes work easier and life happier for all. (5)

While there is no doubt that phone workers had an effect on the shape, uses, and perceptions of the telephone, women at home were also instrumental in using and shaping uses of the telephone. For example, in contrast to industrial histories that characterize the phone's introduction as one of a business technology of primary use to men but with some uses for women (e.g., shopping, personal safety, planning and logistics, sociability), social histories show that men did not take readily to the phone (Fisher 1992; Martin 1991; Rakow 1992). Concerned with loss of privacy (one indication that the telephone is less private than people have come to believe) and the effect that the phone would have on their perceived status within the community, businessmen—especially professionals—tended to relegate phone tasks to secretaries, wives, daughters, and sisters, if they used the phone at all.

Domestic Technology

Because the telephone offered communications capability to the home and was located in the home, it became a domestic technology. And because women were so much in the home, it was a communication tool that they used often, with the result that telephone norms and cultures formed. In fact, women's early extensive use of the telephone for both formal and informal communication with friends and family—what has come to be called "sociability"[7]—had significant effects on its technical and discursive development. Indeed, the telephone was initially marketed as a business tool for men, but Bell Telephone Company found that women tended to use the telephone significantly more for social, interpersonal communication. Women—both rural and urban, modest and elite—called mothers, sisters, other family members, and friends to "keep in touch"; this use comprised such a high percentage of the volume of calls that Bell decided to switch its marketing tactics to push sociability as the primary phone use in the early 1920s. This is an important point. Whereas Bell pushed sociability for users of all genders, this sociability was—and continues to be—culturally ridiculed when applied to women. In fact, as Rakow's (1992) quote here indicates, there is much about women's engagements with the telephone that has been misunderstood and mischaracterized:

> Academic literature has referred to women's telephone talk as gossip, chitchat, and chatter. Authors have claimed, without much apparent sympathy or thought, that lonely and housebound homemakers and farm women found the telephone a solution to their isolation. These characterizations of women and our talk make women's use of the telephone appear to be a consequence of innate differences of responsibility, interests, and personality between women and men rather than an enactment of socially constructed differences resulting from women's assigned place. (2)

What Rakow highlights in this quote is the rosy picture that most telephone scholars have painted of women and the gendered and gender work of talking and listening that they engaged in when using telephones. The reason women were at home in the first place, what this meant for them as members of a larger society, and how it played out in their lives, has been marginalized by these scholars. Mentioned but not adequately explored, for example, are issues such as the phone's role as a strategy for ensuring women's and children's personal safety —a discourse that has resulted in women being cloistered away from public spaces in a variety of ways throughout time. While sociability was certainly an important element of women's interest and use of the telephone, less visible aspects of the story also deserve consideration. Feminist scholars like Rakow (1992) and Martin (1991) have attempted to expose these other stories.

A particularly important but often overlooked theme that these scholars expose is the role that the telephone has played as a primary means for reinscribing gender hierarchies and patriarchal dominations. They claim that phones have foisted more labor and distraction upon women at work and in the home, have tended to enforce women's geographic separation from and movement in spaces outside of the home and office, and allowed the unwelcome intrusions of crank and obscene calls into their private spaces. It is clear, then, that telephones, like any technology, are not power- or person-neutral. Nor can their presence or use be categorically deemed "good" or "bad" as the scholarship represented in my review suggests. While "telephones may help create the [repressive] conditions that they are claimed to mitigate" (Rakow 1992, 32), they are also an extremely accessible technology, one that women use extensively, and one that serves very important organizing and connection functions. Thus, the phone has been contradictorily both a tool of confinement and of connection for women. Within the structural and material constraints of societies, however, social and cultural practices emerge that transgress and transcend—however feebly—both internally and externally imposed constraints. As Wajcman (1991) states: "The telephone has increased women's access to each other and the outside world. In this way the telephone may have improved the quality of women's home lives more than many other domestic technologies" (105). But, as Rakow (1992) opines, the solidarity produced may not have been political in a formal sense:

> If women have distinguished themselves as telephone talkers, as popular mythology has it, one needs to ask why women feel the need to talk on the phone, what work they are doing, what opportunities or limitations give shape to their lives, and what being women means to them and others. The telephone is symptom, possibility, weapon, companion, tool, and lifeline. For the women of

Prospect [the town studied] it has essentially altered the shape of their private lives while a larger world of public participation lies beyond their grasp. It will take their critical awareness of the possibilities for transforming the arrangement of social relations (which is already a point of disagreement) to put the telephone to a different, perhaps political, use. Certainly women in other times and places have done so (154).

Liberal Feminism

Overall, this research emerges from and takes the perspective of second-wave liberal feminism, since "woman" is treated largely as a monolithic gender category which, although class is taken into account, is not substantially explored to account for race. This weakness in the research will be addressed more specifically in the final section of this chapter. However, in this section, I have used this historical information to gesture at how social and cultural historical narratives of women and the telephone map the phone as a gendered, domestic technology that exists in, and spans, the private space of the home. That the phone is used for sociability is widely held and thus discursively furthered by these narratives, although there is disagreement as to its solidarity effects. Those authors who find the telephone to be a tool of solidaristic connection through sociability propose potential spaces of interpersonal activity that, I would argue, inevitably have cultural and political ramifications. Authors like Rakow (1992), Martin (1991), and Moyal (1992) do not argue that the work women do on and through the phone is necessarily, or even primarily, political in a formal sense. However, I am arguing that much of this work is of a political nature.

I would now like to expand this historical narrative with research on the contemporary status of women and phones. While this body of research continues the studies just considered to some degree, I believe that shifts in theoretical approaches to gender, society, and technology problematize and extend what has traditionally been associated with women, phones, and community, and the political solidarities these allow. Most of this research on telephones is grounded primarily in psychological and social-scientific methods, although some studies use postmodern theoretical approaches. Four types of research emerge regarding women and phones specifically:

1) research on the benefit of the phone as an intervention tool in crisis situations; 2) research on the phone as an ongoing networking tool for sociability and support; 3) research on the phone as a wanted or unwanted link to love relationships or sexual/emotional attention; and 4) theoretical work that calls into question telephone technology, its origins, and its uses and users.

Instrumental or Intrinsic

Current social-scientific research on the telephone tends to distinguish phone practices as either instrumental or intrinsic (Nobel 1987; Rubin, Perse, and Barbato 1988). Instrumental uses designate those which further a rational or logistical goal (e.g., making appointments or arrangements, getting or reporting information, intervening in a crisis situation). Pleasurable phoning such as for company, entertainment, or interpersonal interaction comprises intrinsic use. General phone use studies such as Claisse and Rowe's (1993) in France show that two-thirds of phone activity takes place for instrumental uses related to activities and mobility, allowing people to make fewer trips than they would without a phone. In such instances, the telephone extends and makes more public the private domestic sphere and, in doing so, compresses both space and time while also allowing vocal presence to replace physical presence.

Gratification. General uses and gratification studies such as those by Dimmick, Sikand, and Patterson (1994) and O'Keefe and Sulanowski (1995), while categorizing phone use and gratification differently, generally agree that phones are primarily used intrinsically (i.e., sociability, entertainment, and reassurance) and that such use has gendered dimensions. For example, O'Keefe and Sulanowski found that "women use the phone more for social purposes, although their uses may be associated with such factors as fear of crime, confinement to the home by small children, and greater distances separating women from family and friends" (923). Women also use the phone as part of their role as social managers within families. The authors also found that while men and women make a similar number of calls, women talk significantly longer. Forms of gratification associated with social uses are affection, surveillance, and "elements of inclusion and control related to social interaction" (927) (but see also Nobel 1987; Rubin, Perse, and Barbato 1988). These general studies consider gender differences in phone uses and gratification to some degree, albeit in a limited fashion, and tend to extend findings posited in the historical analyses: that women use the phone often; that their use is highly intrinsic; and that they undertake it for a variety of environmental and personal reasons. However, while women's predominance in private space is an essential driver of phone use, this issue is not subjected to analysis in these studies. Nor is race presented as a necessary analytic. Research specific to women's phone use, such as Moyal's below, explores issues of space, geography, and race more pointedly and thus provides more context for the ways and reasons that women engage with the phone and each other.

Kinkeeping. Moyal's (1992) national survey of the phone use of Australian women, perhaps the most comprehensive study to date, largely extends themes from the historical research on women and phones discussed earlier, and provides more specificity. Moyal found that Australian women use the phone primarily to engage in the caregiving labor of "kinkeeping" (maintaining familial relations over both short and long distances), networking between friends, and volunteerism. Unlike the research discussed thus far, Moyal outlines specific use trends for certain demographic segments of women that are highly illuminating and points to the need for further, and more specific, contextualized study of this kind. Moyal (1992) found that for the aged, issues of safety, security, and assistance for independent living were paramount. For immigrant women and women in nonwhite, non-English-speaking, and ethnic communities, the phone proved crucial for both instrumental and intrinsic calling. Access to government information, social services, and jobs as well as to others within the same linguistic or cultural community allowed these women to take care of business and maneuver in Australian society in ways they would be unable to without such a technology. For rural women, Moyal found the telephone to be a "top priority" since it allows contact with loved ones as well as important medical and social services. Aboriginal women find it difficult to get access to the government radio phones housed in outback stations since officials gatekeep access to the phones and usually give men priority. "The solution appeared to several women to be the installation of a second radio-telephone connection, a dual system, that would give ongoing access to Aboriginal women for use in domestic or community violence, in health matters, in seeking information, and for social communication with their children" (303). In this instance of women's desire for private communication in a formal public space, Aboriginal women could not gain access to the technology that would further it. Such a finding highlights the degree to which the telephone serves as a crucial connector of diverse publics in quasi-public spheres.

Support hotlines. Other U.S. and international research corroborates Moyal's findings that the crisis or support hotline is a highly important use of the telephone for women. Whether in the home, on the road, or in a locale of temporary shelter, women around the globe use the phone to contact social support organizations for help, advice, and connection with other women in similar situations (Waters, 1995). Because it is so accessible and allows callers a certain degree of anonymity, the phone is an excellent tool of first response. The telephone helps women break through the confining boundaries of the home or social relations that conceal and reinforce their social and sexual domination, as occurs in situations of domestic violence, sexual abuse, or assault—again, breaking through oppressions that occur in private spaces by allowing connection to quasi-public space.

Phone hotlines and networks also exist for ongoing physical and mental health support. For example, there are a number of phone support networks for women interested in sharing information and advice about AIDS, breast cancer, and other women's health issues such as aging and living alone (Heller, Thompson, and Trueba 1991). While I would argue that hotline and support network phone use is both instrumental and intrinsic to varying degrees, I want to highlight that the talking and listening that women engage in, in these circumstances, is sociable interaction. Certainly calls may, at first, be instrumentally driven. However, even in the most traumatic situations, general sociable communication accompanies and underpins support talk (Moyal 1992). Even in noncrisis situations, this kind of social networking has important instrumental and intrinsic effects. Scott (1996) has found, for example, that women's networks in business and government are crucial in supporting their professional development and advancement. The telephone, of course, is central to forming and maintaining such networks. In these instances, then, the women can transcend private space or enclave[8] space confinements through telephone use that allows them to call out to and create quasi-public spaces or networks of support.

There is a significant difference in calling out and receiving a call, not least of which is the degree of control one has over when and to whom the call is placed. In research that considers the telephone as a site where dating and undesired sex- or power-related interactions take place, women do not typically take the active role in these interactions. For example, Sarch's (1993) study of single women's use of the telephone in dating relationships with men found that women negotiate power, meaning, and representation with males over and through the phone. Expectation is a central theme in this research. Women not only expect and wait in private spaces for telephone calls from men; they also have specific expectations of their own behavior regarding such calls (available, amiable, passive, etc.). This is behavior that is both expected by men and by society more generally. One of these expectations is that women should not be the active party in phone communications. Martin (1991) notes that historically, women have been dissuaded from initiating romance-related calls. And this tradition of men being phone communication initiators who are able to invade private space may play into sexual/power games such as men collecting women's phone numbers without the intention of calling—a game more of pursuit and intrusion than equitable or even mutual interaction (Edwards 1996). Sarch's study suggests that these kinds of power games and relations exist in society and are reproduced on the phone. In the case of dating, then, these studies find that the telephone seems to continue to be a technology of containment, constraint, and oppression for women.

Intrusion

Oppressive as well are obscene and harassing phone calls that women receive in their homes and at work through unsolicited, usually sexual, attention by men. In these cases, phoning becomes an invasion or intrusion which demonstrates the significant power differentials that occur between dominants and non-dominants in even quasi-public, technical spaces like the space of the phone connection. According to Katz (1994), Smith and Morra (1994), and Shefield (1989), on average women have between a 65.3% and 90% chance of receiving obscene or harassing calls during their lifetimes. Women do not seem to place a great deal of faith in police or telephone company responses to these calls, which typically place the burden of intervention back onto the women themselves. Thus reporting of this kind is low (35%). Usually, women try to mitigate the possibility of receiving such calls by giving their phone number only to people they know to be reputable, taking an unpublished number, or listing their number in the phone book under their husband's name or only with a first initial. However, technological solutions are also employed, which recent advances in telephone-related technologies have made possible: screening answering machines, voice mail, and caller ID allow women greater control over answering incoming calls as well as exercising some level of surveillance. Even these strategies are no assurance against phone intrusions, however. Relations of domination, particularly sexual domination, continue to be replicated through phone technology. While these researchers do not address women's strategies for coping with the trauma of telephone intrusions, my experience is that women engage in therapy and talk with one another to assuage fear, receive support, and strategize ways of avoiding further incidences.

Contemporary research supports the notion that women use telephones to forge community by extending their private spaces to quasi-public space. It demonstrates that when the directionality and the caller's gender are changed, the power dynamics and safety of quasi-public space change, unfavorably, as well. What the social, historical, and contemporary research fails to consider, however, are those instances where women use the phone actively to enter not-women-only, quasi-public spaces for particular instrumental and intrinsic purposes. In the next section I consider a few such instances and some of the more theoretical scholarship that, I believe, draws upon the experiences and possibilities of such scenarios.

Expanding the Counternarrative of Women and Phones

In the feminist historical and social-scientific research I have presented so far, sex and gender are assumed and essentialized by both.

In the case of the former scholars, I would argue that the tendency to essentialize gender is the result of their emerging from a largely second-wave feminist tradition (e.g., Martin 1991 and Rakow 1992, clearly more Marxist; and Moyal 1992, more liberal). In the case of the latter, conventions of quantitative socio- and psychometric methods insist that valid and reliable categories such as male/female gender be used. This may be more a matter of theoretical and historical timing than anything else. For example, only recently has theory asserting nonessentialist sex and gender categories, such as that by Butler (1990, 1993) on gender performativity in constructions of sex and gender identity, begun to be taken seriously outside of feminist circles. And in the case of technology, it is becoming increasingly accepted that technologies and the forms they take are social constructions over which humans do or do not have material or theoretical control. On this account, Haraway's (1989, 1991, 1997) and Latour's (1993) work on the cyborg and the quasi-object respectively, which calls both humanness and technology into question, requires us to rethink identity and subjectivity as distinctly nonhumanist.

In contrast to what the research outlined in the section above would suggest, women's sexual domination through the telephone is neither a necessary nor a totalizing occurrence. Indeed, the discourse of phones as tools of gender and sexual oppression is weakening as women themselves begin to engage more in obscene and harassing phone behavior (or as extant behavior begins to be perceived) (Gill 1995), and as consideration of the complexities of overt sexual uses of the telephone become more visible. Within the psychological literature there is a body of research that explores male and female "compulsive" and "perverse" telephone use. This literature argues, predominantly, that voice and the telephone can be symbolically construed as phallic and vaginal extensions, with the powers and passivities traditionally associated with each. The telephone can be used as a fetish and, when used in this way, it is usually to assuage feelings of loneliness and abandonment or to express power through intrusion by the phones as fantasy phallus.[9] Richards (1989), for example, argues that women have long exhibited perversion and expressed it through media such as the telephone. However, because perversion has been largely understood as a male phenomenon, these kinds of behavioral expressions have been under-studied and misdiagnosed.

When one thinks of the phone as fantasy phallus and pays attention to cases where it has been a conduit of pleasure and pain, one can begin to see the telephone's utility in allowing complex expressions of identity and power within a society that highly constrains women's expression of sexuality and aggression. These kinds of expressions can also be found in the research on other sexual uses of the phone.

Research on phone-sex workers and the phone-sex industry shows that while traditional relations of gender domination do exist, they are not absolute.[10] While the telephone can extend traditional notions of sexual pleasure, sexual space and, thus, sexual power relations, phone-sex workers are able to transgress and shift these traditional power relations largely because their voices, not their bodies, give erotic pleasure. With disembodied sexual acts as the stuff of trade, sex workers can avoid the dangers of embodied sex work (Cybersex: An adult affair, 1997; Stone 1995). And this pleasure of voice and disembodied sex is not limited to heterosexual encounters or paid work. For example, in Hanson's (1995) piece on the use of the telephone by gay men with HIV or AIDS to express mourning and paranoia, he emphasizes the phone's importance to gay/queer communities in providing a venue for connection, mourning, fantasy, and so forth as they create, perform, and maintain group and individual pleasure (often sexual) in the face of the new, individuating and alienating practices that HIV/AIDS has imposed:

> Phone sex may be an act of mourning for an idealized sexual freedom rumored to have not disappeared; on the other hand, it may be a refusal to mourn and a challenge to the validity of the loss itself. The telephone calls up an electrical space that becomes a queer space, a new space of sexual play and sexual imagination. (45)

Hanson's research challenges the claims that the social-historical and social-scientific research explored here make about women maintaining a monopoly on intrinsic uses of the phone. Hanson's counter-counternarrative is that men (here, gay and queer men) use the phone for creating community around deeply emotional purposes such as mourning. He also makes clear (by association, not explication) the relational and community-building salience of space-spanning communication technologies for all nondominant members of society who experience hostility or marginalization in dominant public spaces. Ultimately, Hanson begins to blur the lines about telephone technology, arguing that it is "queer" because its technology both manifests and ambivalently transcends the binaries of distance/intimacy, defense/vulnerability, control/uncontrol, pleasure/terror, perversion/normalcy, embodiment/disembodiment, and so on. Of course, this claim is also made of the Internet, and has been made of most emerging communications technologies—albeit using different metaphors.

Phone and Identity

What this literature does is make both the telephone and its users strange. This "maddening" or "queering" of the identity of the phone user and of the phone itself as a space of meaning draws attention to—and calls into question—the entire project of this chapter. Can one talk about "women" and the "telephone" in bald terms and accept that they exist in the particular communicational spaces (private) attributed to them, unproblematically, thus far in the research? As Hanson (1995) and Stone (1995) suggest, the terms of the discourse of telephones have been heavily influenced by the theoretical assumptions that inform them. What happens if one considers research about the telephone that adopts postmodern philosophical and psychoanalytic approaches to the telephone? One scholar, Ronell (1989), construes the telephone as a trope of terror among philosophers and social theorists; schizophrenics and spiritualists, inventors, and vaudevillian performer as well as those with such diverse and conflicting "social ameliorative" projects as helping the deaf hear or politically oppressive projects such as Nazism. She asks if schizophrenia has replaced philosophy in our time. For, like philosophers (e.g., Heidegger), schizophrenics have long "answered the call" of unworldly, ungrounded voices which provide distinct, albeit distinctly different, knowledges and apprehensions of life. Her research highlights the degree to which the telephone's technical development was strongly tied to spiritualists and quasi-vaudevillian performers such as Alexander Graham Bell and Thomas Watson, who sought to use telephony for either actually communicating with the underworld or, barring that, taking the illusion (but still the possibility) on the road. Ronell poses another question: Can a person be sure that schizophrenics, philosophers, and spiritualists are indeed not human telephones? Can one really rest assured that the telephone rationally does what people think it does (i.e., connect two embodied people by electric/electronic means)? When I hear what my friend calls "martians" on the phone line—crackling and screaming high- and low-pitched gibberish over fluctuating frequencies—I assure myself it is "nothing but interference." Yet, interference takes many forms, not all of them known or knowable. Ronell's work highlights often unasked but certainly unanswered questions: What kind of interference? Who or what is on the line? Her theoretical and empirical approach shows that central is the terror of the unknown, the uncontrollable that exists centrally within embodied technological and human day-to-day experience, but which has been normalized to the point of vanishing.

Expanding Subjectivity

This research on the telephone points to the undeniable and unignorable challenge that scholars like Butler (1990, 1993), Haraway (1989, 1991, 1997), Stone (1995), and Ronell (1989) put to widely held ontological and epistemological assumptions about identity, technology, and space that are dominant to date. Who or what is a normal human without an extensive and extending technology of communication through which it can perform itself? What is a communication technology without its human underpinnings and infrastructures? And, do these indeterminacies destabilize and reconfigure the categories of space and politics that scholars have come to rely upon to explain human interaction using technology? Oddly enough, it is this literature on the phone's queerness and perverse uses that makes the phone strange enough to begin to see glimpses of the distinct possibilities it has always held for communication and individual expression in service of interpersonal relations and community building; yet this should not be surprising. For, as with any process of "making strange," the closer an intervention comes to taken-for-granted categories and practices, the more it calls these categories and practices into question.[11] Gender, sexuality, sexual behavior, and telephoning are all highly reified and normalized categories which command expectations of particular practices. I begin to see ruptures in these, however, in the literature outlined here. The feminist/social historical work that minimizes the work and relations women have undertaken so persistently and creatively using the telephone does not seriously address the use of phones for social or political organizing, largely ignores phone use by nonwhite women, and does not consider to any significant degree the ways that phones can be used to bend identity and identity-related practices.

Solidarity

What the research outlined above suggests is that the technology of the telephone allows "subtle" and "supple" solidarities: very complex, flexible, situational connections that are made and broken as desire and need require. I see them as an approximation of Haraway's (1989) notion of "affiliation." I also see them as productive, hybrid locations of culture such as Bhabha (1990) propounds. For while they are fluid and ever in process, they are so in relation to environmental forces and constraints. Contrast this with the hard and fast solidarity of the bourgeois public sphere, one limited to rational discussion, to a particular public space, to formal politics accessible to certain people in certain times and places. Of course, the subtle solidarities of which I speak occur even in these forums, since they are part of the more formal communication that occurs—indeed, they may be more constitutive of it than people currently suspect.

Gendered space? Communication, communicational, and relational space are central elements of the human experience. While they are not essentially gendered, there are gendered or nondominant "tendencies" of expression and practice due to social policing of identity, behavior, and technology—many of which encourage situational expressions and practices or subtle and supple solidarities for nondominant groups. Thus, that women use the telephone as a technology, which allows for the creation and use of divergent public or quasi-public spheres, is significant. However, research on telephone use has been so limited and so traditionally and unproblematically category-bound that scholars have yet to ask questions that shed significant, divergent, maddened, and queered light on what these particular technical spaces are; what not-traditionally-thought-of communications and solidarities occur within them; and thus, what they mean in and of themselves and in relation to society. This work needs to be undertaken in relation to the phone as well as other communication technologies. However, as I have tried to demonstrate, dominant narratives of the space and use of the telephone continue to advance more modernist notions of identity, communication, political action, and solidarity. In the end, these notions may not be all that discursively or materially salient for political projects of unfettered expression, equitably (or at least, mutually acceptable) affiliational communication, and, through these, further social, political, economic, and cultural transformations.

What is interesting, and potentially more useful, is the construction of quasi-public spheres. They are locales which not only formulate emergent political action, but express fully formed political utterances and acts. Unfortunately, most of the telephone scholarship considered in this chapter privileges public political action through unidirectional movement from private to public populations and spaces. That it does so would seem to suggest that the assumptions upon which it is based are inadequate for categorizing and understanding current social forces and relations. More useful, perhaps, are approaches that value the power and possibility of the nondominant (see de Certeau 1984). These approaches make it theoretically possible to consider that the private might colonize, destabilize, or otherwise transform the public and what has traditionally been associated and validated within it.

Thus I come full circle. Is the quasi-public space of the telephone a site of political possibility? Certainly it is, but not merely in the way that the liberal feminist, social scientific, and psychological literature has narrated it. Is it a site of exclusion, containment, or oppression? Yes, it is this as well. However, to understand better how these two ends of the continuum work and the myriad of possibilities in between, scholars need to better understand the day-to-day practices that people engage in and the meanings they make of them. De Certeau (1984) provides insight:

The presence and circulation of a representation (taught by preachers, educators, and populizers as the key to socioeconomic [or political] advancement) tells us nothing about what it is for its users. We must first analyze its manipulation by users who are not its makers. Only then can we gauge the difference or similarity between the production of the images [narration] and the secondary production hidden in the process of its utilization. (xiii)

Although the telephone is a natural and normalized technology discourse in U.S. society, considering it more closely, and in ways that try to destabilize what has become discursively stable, is one way to tactically consider what discursive, technical, social, and political power lays dormant, or latent, within a scholarly reach.

Notes

1. The term "solidarity" and the philosophical and practical positions that it implies have become central in discourses of democracy —an expansion from its origins in Marxist-inspired class struggle. While the class basis of solidarity in the struggle against capitalist systems of political and economic oppression has both eroded and expanded, what has remained salient in the concept is the importance of collective political action against a dominating force or forces.

2. I use the term "women" as an analytic category in this piece with hesitation, for it is not unproblematic. Not only has the construct "women" been subject to important critique (Butler 1993; de Beauvoir 1953; Haraway 1989), its performance, determination, and mobilization become even more problematic in telephone spaces because identity, including gender identity, can be manipulated. "Women," then, is used throughout this chapter as a heuristic. It is a term employed to variously signify "women" in the normative sense, and a more general kind of nondominant subject position that could, and often does, apply to other nondominant subjects as well. The text's context should make the different intended emphases clear.

3. I use the term "broadcast" here in its general sense, not the technical sense as defined by the FCC. Thus, "broadcast" means to cast, scatter, or distribute a message widely.

4. Kant conceived of the public sphere as spaces and organizations where groups of private citizens—typically upper-class males—engaged in reasoned public debate about political issues. Thus, they displaced, or at least destabilized, the monopoly of formal public authorities on public decision making.

5. Habermas has modified his conceptions based on numerous critiques and has moved most notably toward his "theory of communicative action." See his "Further Reflections on the Public Sphere" in Calhoun (1992).

6. See Maddox (1977), Green (1995), Lipartito (1994), and Norwood (1990).

7. Telephone sociability is not a very well-defined analytic concept. Social historical research, as described here, tends to adopt the phone company's early definition: general social activity on the phone (see Fisher 1992; Rakow 1992). More recent studies have defined sociability as an intrinsically motivated behavior for receiving gratifications of pleasure, affection, inclusion, and relaxation (O'Keefe and Sulanowski 1995, 931).

8. I use the concept "enclave" here in the sense of a spatial cloistering for the purpose of garnering and imposing power. See Fiske (1994).

9. See also Almansi (1985).

10. That women, particularly poorer women and women in nonindustrialized countries, engage in sex work by force of economy or threat to personal well-being is well documented and continues to call into question social, political, economic, and cultural values about women and their work. However, some women choose to engage in sex work willingly. Either way, the telephone mitigates the dangers of embodied work of this kind to some degree.

11. Consider this quote by ethnomethodologist Harold Garfinkel (1967) on the analytic opportunities afforded by close consideration of the experiences of hermaphroditic subjects: "The experiences of ... intersexed persons permits an appreciation of ... background relevances that are otherwise easily overlooked or difficult to grasp because of their routinized character and because they are so embedded in a background of relevances that are simply 'there' and taken for granted" (118).

References

Almansi, R. 1985. On telephoning, compulsive telephoning, and perverse telephoning. *Psychoanalytic Study Society*, 11:217-235.

Bhabha, H. 1990. *Nation and narration*. London: Routledge.

Butler, J. 1990. *Gender Trouble: Feminism and the subversion of identity*. New York: Routledge.

———. 1993. *Bodies that matter: On the discursive limits of "sex."* New York: Routledge.

Calhoun, C., Ed. 1992. *Habermas and the public sphere*. Cambridge, MA: MIT Press.

Cybersex: An adult affair. 1997. *The Economist* January 4: 64-66.

de Beauvoir, S. 1953. *The second sex*. London: Penguin.

de Certeau, M. 1984. *The practice of everyday life*. Berkeley: University of California Press.

de Sola Pool, I. 1977. *The social impact of the telephone*. Cambridge, MA: MIT Press.

Dimmick, J., Sikand, J., and Patterson, S. 1994. The gratifications of the household telephone: Sociability, instrumentality, and reassurance. *Communication Research* 21(5): 643-663.

Edwards, W. 1996. Call waiting. *Essence* 27: 57-8.

Federal Communications Commission-Common Carrier Bureau. 1998. FCC releases new telephone subscribership report. Washington, DC.

FIND/SVP Emerging Technologies Research Group. 1997. *1997 American Internet user survey*.

Fisher, C. S. 1992. *America calling: A social history of the telephone to 1940*. Berkeley: University of California Press.

Fiske, J. 1994. *Media matters*. Minneapolis: University of Minnesota Press.

Fraser, N. 1992. Rethinking the public sphere: A contribution to the critique of actually existing democracy. In *The Phantom Public Sphere*, edited by B. Robbins. Minneapolis: University of Minnesota Press.

Garfinkel, H. 1967. *Studies in ethnomethodology*. Cambridge, U.K.: Polity Press.

Gill, M. S. 1995. The phone stalkers. *Ladies Home Journal* 112:82-87.

Green, V. 1995. Race and technology: African American women in the Bell system, 1945-1980. *Technology and Culture* 36(supp): S101-43.

Hanson, E. 1995. The telephone and its queerness. In *Cruising the performative: Interventions into the representation of ethnicity, nationality, and sexuality,* edited by E. A. Case. Bloomington: Indiana University Press.

Haraway, D. 1989. A manifesto for cyborgs: Science, technology, and socialist feminism in the 1980s. In *Coming to terms: Feminism, theory, politics* edited by E. Ied. New York: Routledge.

———. 1991. The actors are cyborg, nature is coyote, and the geography is elsewhere: Postscript to "Cyborgs at large." In *Technoculture,* edited by C. P. A. Ross. Minneapolis: University of Minnesota Press.

———. 1997. *Modest_witness@Second_millennium. FemaleMan_meets_oncomouse.* New York: Routledge.

Heller, K., Thompson, M., and Trueba, P. 1991. Peer support telephone dyads for elderly women: Was this the wrong intervention? *American Journal of Community Psychology* 19:53-146.

Katz, J. E. 1994. Empirical and theoretical dimensions of obscene phone calls to women in the United States. *Human Communication Research* 21:155-82.

Langer, E. 1972. Inside the New York Telephone Co. In *Women at Work* edited by W. L. O'Neill. Chicago: Quadrangle Books.

Latour, B. 1993. *We have never been modern.* Cambridge, MA: Harvard University Press.

Lipartito, K. 1994. When women are switches: Technology, work, and gender in the telephone industry, 1890-1920. *American Historical Review* 99 (Oct.):1075-1111.

Maddox, B. 1977. Women and the switchboard. In *The social impact of the telephone* edited by I. D. S. Pool. Cambridge, MA: MIT Press.

Martin, M. 1991. *"Hello, Central?": Gender, technology, and culture in the formation of telephone systems.* Montreal: McGill-Queen's University Press.

Morley, D. 1986. *Family television: Cultural power and domestic leisure.* London: Comedia.

Moyal, A. 1992. The gendered use of the telephone: An Australian case study. *Media, Culture, and Society* 14:51-72.

New York Telephone Company. ([n.d.] 1977). An ideal occupation for young women [Microform]. History of women, no. 8604. Woodbridge, CT: Research Publications.

Nobel, G. 1987. Discriminating between the intrinsic and instrumental domestic telephone user. *Australian Journal of Communication* 11:63-85.

Norwood, S. H. 1990. *Labor's flaming youth: Telephone operators and worker militancy, 1878-1923.* Urbana: University of Illinois Press.

O'Keefe, G., and Sulanowski, B. 1995. More than just talk: Uses, gratifications, and the telephone. *Journalism and Mass Communications Quarterly* 72(4):922-933.

Rakow, L. 1992. *Gender on the line: Women, the telephone, and community life.* Urbana: University of Illinois Press.

Richards, A. K. 1989. A romance with pain: A telephone perversion in a woman? *International Journal of Psycho-Analysis* 70:153-164.

Ronell, A. 1989. *The telephone book: Technology, schizophrenia, electric speech.* Lincoln: University of Nebraska Press.

Rubin, R., Perse, E., and Barbato, C. 1988. Conceptualization and measurement of interpersonal communication motives. *Human Communication Research* 14(Summer): 602-628.

Sarch, A. 1993. Making the connection: Single women's use of the telephone in dating relationships with men. *Journal of Communication* 43:128-44.

Scott, D. 1996. Shattering the instrumental-expressive myth: The power of women's networks in corporate-government affairs. *Gender and Society* 10:232-47.

Shefield, C. 1989. The invisible intruder: Women's experiences of obscene phone calls. *Gender and Society* 3(4):483-488.

Smith, M., and Morra, N. 1994. Obscene and threatening telephone calls to women: Data from a Canadian national survey. *Gender and Society* 8:584-96.Spender, D. 1995. *Nattering on the Net.* North Melbourne, Victoria, Australia: Spinifex Press Pty. Ltd.

Stone, A. R. 1995. *The war of desire and technology at the close of the mechanical age.* Cambridge, MA: MIT Press.

Wajcman, J. 1991. *Feminism confronts technology.* University Park: Pennsylvania State University Press.

Waters, R. 1995. A hotline movement grows in Russia. *Ms.* 6:19-21.

Chapter Six

Chinese Online Presence: Tiananmen Square and Beyond

Yan Ma

The riotous bloom of people power, Chinese style, that took hold of Beijing last week began as a movement almost exclusively of students. But in one of those extraordinarily rare and historic occasions—it was Karl Marx who gave such moments the classical definition "revolutionary praxis"—a kind of instant solidarity appeared. ... It bound together the disparate groups—students, workers, professionals, academics. ...

When it happened, suddenly a million or more marchers were streaming into Tiananmen, perhaps ten times as many as had been there the day before. It was the largest demonstration in modern Chinese history. People poured out of factories and hospitals, the Foreign Ministry and kindergartens. And not just in Beijing. By midweek the ferment had spread to at least a dozen other cities, with another hunger strike taking place in Shanghai. In some provincial cities plans for a general strike were reported.

Benjamin 1989, 40

By the time the world witnessed the unfolding of a massive but traditional resistance strategy in Tiananmen Square, Chinese dissidents had established a vast online presence. In fact, as the Zapatistas and U.S. environmentalists were creating Web sites for the Indians in Chiapas, Mexico, during the late eighties, Chinese students and other proclaimed Chinese democrats were faxing and emailing their petitions around the world. As members of a secret nongovernmental organization, Chinese dissidents understood the reach of decentralized electronic communications and structured their organization to include Chinese expatriates, Chinese Americans, Chinese Europeans, and democratic sympathsizers around the world. Forbidden an actual presence under the Chinese state, they coalesced into a virtual democratic citizen-group online.

Like the Zapatista construction of the Chiapas Indians' plight, Chinese dissidents suing for particular civil rights such as freedom of speech, assembly, and the press, structured their appeals as a democratic pursuit. Unlike the Zapatistas, who have a Marxist background, these dissidents were fighting against Communist ideology and were enamored of democracy. Chinese student dissidents seeking "a clean sweep of China's rampant corruption ... [and] a free press" (Benjamin 1989, 44) spoke to U.S. and other Western citizens using the rhetoric of democracy. They used or paraphrased slogans familiar to Americans such as "I Have a Dream" (or) "Give Me Democracy or Give Me Death" (Benjamin 1989, 40). In a more thoughtful manner, they quoted Abraham Lincoln and Thomas Jefferson, but their ideas of democracy were as yet underdeveloped. Nonetheless, they used demassed communication technologies in effective ways to transmit fledgling ideas of democracy. They approached individual rights with a fervor born of years of suppression of the individual in favor of the collective. With this sincere quest for personal rights, they petitioned world citizens, especially those in the United States, who embraced democratic ideology. By faxing, emailing, and creating Web sites, Chinese dissidents effectively siphoned off some of the control of the flow of information from their Communist government and augmented their voice and power by soliciting allies around the world.

Chinese dissidents understood the power of a virtual organization, a virtual presence when their actual presence was and is so constrained within China. I am interested not only in the manner in which this happened, but also in the way that communication technologies have been used to promote the welfare of ordinary citizens in our time and earlier. In this chapter, I explore the Tiananmen Square incident and the manner in which ordinary citizens used fax and phone to communicate their demands and plights. I describe the social role of other demassed technologies and note their use in contemporary resistance efforts. Finally, I present the results of an informal survey I conducted on the use of fax for disseminating information during the Tiananmen Square incident.

Tiananmen Square

A decade later, in the wake of many successful democratic resistance movements, U.S. citizens may forget the massive demonstration that was the largest protest in twentieth-century China's history. In the spring of 1989, Chinese students used the occasion of a memorial service for the late Hu Yaobang, a reformer among the top Chinese Communist leaders, to gather in Tiananmen Square. Their demands focused on free expression and an end to corruption in the Chinese Communist party, but went unheeded by government leaders. Citizens joined students in the square by the thousands and began a hunger strike. At the opening of the protest, the Liberal Party leader Zhao Ziyang still had control of the media. Several Chinese newspapers and television stations were allowed to support the protests. However, troops stationed outside of Beijing were ordered to Tiananmen Square to contain the protestors. Beijing dissidents swarmed the soldiers, sitting on their tanks, offering them food, and reminding them of their bonds to the people. All of this took place as the world watched, but Chinese government officials responded to dissident demands by shutting down all satellite dishes operated by foreign television networks; horrified global viewers feared for the worst (Benjamin 1989). One avenue of information had been blocked.

Faxing, which was one link the dissidents had established with Chinese students and Chinese-Americans abroad, now assumed major proportions. Fax machines had been widely installed in Chinese universities, hotels, and international business offices. Chinese students in the United States collected about 1,500 fax numbers and posted them on computer bulletin boards; various groups began blindly faxing news reports of the revolt to recipients at those phone numbers in the United States, including those working at Chinese American newspapers. "[This was] the electronic equivalent of a note in a bottle. In China, students, hotel waiters or office workers retrieved the messages; then they were reproduced by the hundreds in photocopiers and put on public display" (Martz 1989, 29).

The manner in which information was transmitted to and from Beijing during and after this event is important and presents intriguing considerations of the role of faxes in reporting news. During the Tiananmen Square incident, fax machines were used to their fullest potential in disseminating information and bypassing censorship. News was successfully transmitted to countries beyond China's borders. "Dissidents, then, used decentralized communications to send their news to Chinese students and scholars around the world, and it was these people outside China who continued to fax published items from foreign newspapers back to the dissidents. It was estimated that 30,000 fax machines existed in China at that time" (Martz 1989). Nonetheless, the student movement was crushed by the Chinese militia on June 4, 1989.

Communication Technologies and Citizens

The relationship of decentralized technology to political rebellion is as old as Martin Luther and as new as the presence of electronic *samizdat* and *magnitizdat* (underground videocassettes) in the former Soviet Union. Chinese and other political dissidents often act in opposition to the mass production of messages by the state in this new information age. It is access to decentralized technologies that enables the voices of political resisters to be heard. Historically, only two other decentralized and personal technological innovations, books and telephones, had the rapid rate of diffusion equal to that of today's electronic communications, Books were a product of a premodern information revolution; they too changed social and cultural structures once they reached the hands of ordinary citizens (McLuhan 1964).

Decentralized Technologies

The promises that flowed with the exponential growth of technology in the nineteenth and twentieth centuries are part of the political heritage of many nations, but those innovations, books and phones, to which most citizens had access, changed lifestyles, educational preparation, job prospects and even architecture (McLuhan 1964).

Books. Books were a pre-Enlightenment innovation, but the invention of paper for books was ancient and type printing was an early medieval discovery. Paper was invented in China as early as 200 B.C.; movable type printing was invented by Bi Sheng in 1041-49 (Pacey 1990). A technological revolution did not occur, however, because there was little demand for printed information in eleventh-century Imperial China. The Chinese used full-page wood blocks for printing small numbers of copies. But when these inventions were introduced to the West they were placed in full use. "Between 1481 and 1501, 268 printers in Venice turned out two million volumes" (de Sola Pool 1990, 4). In 1992, the United States alone produced 49,276 titles (*Bowker Annual Book Trade Almanac* 1994). And the printing revolution gave rise to occupations such as publishers, editors, journalists, librarians, and workers in related professions.

Phones. Phones appeared in the second wave of the industrial revolution in the United States and started a communications or information revolution that spurred changes not seen on such a large scale since the initial industrial revolution. The information revolution enabled dissidents in China and elsewhere to spread their messages. In 1980, each person in the United States made approximately 1,000 telephone calls per year (Pierce 1977, 166). And according to the U.S. Bureau of Census (1991), 107.8 million overseas telephone calls were made in 1989.

Assessing Decentralized Communication Technologies

It is difficult to assess the benefits of technological innovations for their ability to extend the voices of ordinary citizens to counteract the dominant reach of state and corporate interests. Beyond utopian claims spurred by the vast reach of modern technologies lie thoughtful claims about the impact of telephones and computers on political resisters around the world. Ithiel de Sola Pool (1977), describing the telephone as liberating, notes that the phone, while protecting privacy, makes social and personal information available. And phones affect business circles as well as urban and rural lives. The phone, he warns, adds to human freedom, but those who gain that freedom can use it however they choose.

In 1983, de Sola Pool observed that civil liberties function in a changing technological context and that the causal relationship between technology and culture has long been debated by social scientists. "Freedom is fostered when the means of communication are dispersed, decentralized, and easily available, as are printing presses or microcomputers" (de SolaPool 1983, 5). His definition of "new communications technologies" includes cable TV, video recorders and discs, satellites, facsimile machines, computer networks, computer information processing, digital switches, optical fibers, lasers, electrostatic reproduction, large-screen and high-definition television, mobile telephones, and new methods of printing. With new communication technologies, distance as a barrier is reduced; the separation between speech, text, and pictures diminishes, and information processing becomes part of daily human activity. Furthermore, electronic communication is global, and computer technology becomes more sophisticated in providing information for specialized use. He believes that electronic technology will be conducive to freedom: "The world is getting smaller because of the availability and accessibility of all these new technologies. National boundaries will play a lesser role in the age of global electronic communications" (1983, 231).

Bill Nichols, a scholar who writes about cyberspace, follows the lead of philosopher Walter Benjamin in his prescient discussion of technology in "The Art of Reproduction in the Mechanical Age" (Benjamin [1936] 1986). Titling his essay, "The Work of Culture in the Age of Cybernetic Systems," Nichols (1988) assesses the human impact of contemporary computer systems. Cybernetic systems, he reminds the reader, include a range of computational systems, such as telephone networks, communication satellites, and laser videodiscs. As Benjamin indicated, the authenticity of an artist's work is lost during the process of reproduction. The same is true in cybernetic communication:

> The chip replaces the copy. Just as the mechanical reproduction of copies revealed the power of industrial capitalism to recognize and resemble the world around us, rendering it as commodity art, the

automated intelligence of chips reveals the power of post-industrial capitalism to simulate and replace the world around us, rendering not only that exterior realm but also interior ones of consciousness, intelligence, thought and intersubjectivity as commodity experience. ... The transgressive and liberating potential which ... Benjamin found in the potential of mechanically produced works of art persists in yet another form. The cybernetic metaphor contains the germ of an enhanced future inside a prevailing model that substitutes part for whole, simulation for real, cyborg for human, conscious purpose for the decentred goal seeking for the totality—system plus environment. (Nichols 1988, 33, 46)

Where de Sola Pool recognized liberation in demassed technologies, both Benjamin and Nichols locate the "transgressive and liberating potential" in new electronic forms. The ambiguity of a cybersystem in which a part stands for a whole but which system allows disembodied experiences must be carefully weighed. These and other scholars, however, were unable to predict the manner in which contemporary decentralized technologies would be used for political resistance in the 1980s and 1990s.

Decentralized Technology and Resistance

In the late 1980s, email, computer networks, fax machines, cable news networks, short-wave radios, and similar devices all played important roles in the revolutions of Eastern Europe and China. East Germans learned about large antigovernment demonstrations through West German TV transmission. In the same way, they found out when Hungary opened its borders to East German refugees and where cracks were opening in the Berlin Wall. During the Tiananmen Square incident, students in Beijing used satellites, fax machines, hand-held TV cameras, computer printers, copiers, and global communication networks to spread information about the progress of that event. Students in China and the United States exchanged information through fax. "There were 2.5 million FAX machines in the United States in 1989, churning out billions of pages of FAXed documents per year. ... Like phones and VCRs, FAXes will begin to appear in even the humblest homes, driven by the Law of Ubiquity" (Toffler 1990, 363-64).

Martz (1989, 29) provides a summary of a survey of decentralized Chinese technologies that he called technologies of freedom:

- Letters delivered in China, 1978: 2.8 billion
- Letters delivered in China, 1988: 6 billion
- Urban telephone subscribers in China, 1978: 1.2 million
- Urban telephone subscribers in China, 1988: 3.6 million
- Long-distance telephone calls handled by the Chinese telephone services, 1978: 185.7 million
- Long-distance telephone calls handled by the Chinese telephone services, 1988: 646.2 million
- The number of long-distance telephone lines from Beijing to provincial capitals, 1980: 27
- The number of long-distance telephone lines from Beijing to provinces, 1988: 89
- Television sets in China, 1978: 3 million
- Television sets in China, 1987: 116 million
- Facsimile machines in Chinese government post offices, 1978: fewer than two dozen (est.)
- Facsimile machines in Chinese public buildings, 1987: 1,410

The technologies of dissemination that the Chinese state considered so important for progress set the scene for coalescing nongovernmental organizations.

Fax and American Chinese-language Newspapers

To examine how fax played a role in gathering and disseminating information about the Tiananmen Square incident, I contacted the editors of thirty major Chinese-language daily, weekly, and monthly newspapers published in the United States to determine the use of fax in their daily newsgathering and reportage of the incident. I conducted informal telephone interviews with fourteen of the thirty editors and sent surveys to the others. Five editors returned the survey. Thus, nineteen publications were represented, including thirteen dailies, five weeklies (including one published in English) and two monthlies.

I selected Chinese-American newspapers printing in the Chinese language to examine the use of fax for several reasons. First, when information came from China, there was no need to translate, so its authenticity was maintained. Second, the Chinese language newspapers are easy for Chinese and Chinese-Americans to read. In the United States, they are the major information resource on China and the Chinese.

In these interviews and surveys, I was interested in the relative importance of faxing during the Tiananmen Square protest compared with its importance in more typical newsgathering. In every case, faxing increased in importance during the protest for many of the reasons cited below. But, in addition, faxes assumed more importance precisely because of the Chinese crackdown that closed off various sources of news. What follows is an informal summary of the responses and comments about the role of faxing as related by the editors of these nineteen publications.

- All the editors used fax machines to receive news. The percentage of news received by fax about the Tiananmen Square incident ranged from twenty to ninety.

- Editors received news on the Tiananmen Square incident from China, France, Hong Kong, Taiwan, and sites within the United States. After the Tiananmen Square incident, the situation changed and no faxed news came from China except from the official Xin Hua News Agency. News, however, could still be faxed to China.

- Editors indicated the advantages of faxing primarily to be the timeliness of the news; that materials could be sent and received any time of the day or night; the lower cost and higher speed of faxing compared to other means of gathering and disseminating news; and that items could be sent hand-written.

- Faxed materials in Chinese were the main source of information for publication on the progress of the Tiananmen Square incident.

One English-language weekly did not receive a single piece of faxed news from China at that time because, according to the editor, it would have taken too much time to translate the Chinese version into English before the news could go to press. On the other hand, another Chinese language weekly indicated that it received faxed news every hour or two during the incident.

- Dailies depended more on fax or satellite than did weeklies or monthlies. This is consistent with the editors' comments that faxes were more timely and accessible.

- All editors believed that fax and technology played an important role in news gathering and reportage on the Tiananmen Square incident.

It is important to note that while faxing was an important component of news gathering and dissemination as related by these editors, these news outlets also rely on reporters (in various countries, often exile dissidents), wire service reports, phone, and even CNN for their news. The point is that faxing became the primary source of reporting on internal events related to the Tiananmen Square protests at the time when these other outlets were being restricted by the Chinese hard-liners.

Russia

In 1991 there were 200,000 fax machines in the Soviet Union (Berkowitz and Quittner 1991), and not surprisingly, fax also played an important role in the defeat of the attempted Soviet coup. In an effort to form new leadership, Soviet hard-liners tried to seize power from the moderate president, Mikhail Gorbachev, who was on vacation in his Crimean dacha. The first news of the Soviet coup came to the United States through a fax machine at the Center for Democracy in Washington at 10:54 p.m. EDT, August 30, 1991. According to PR Newswire, it read: "It is a Coup. Tanks everywhere." Faxed messages continued to come into the center for the next two days. Not only did the first words about the coup come to the United States through a fax machine, but information from the West went to the Soviet Union through the same technology. Using fax, the Russians asked to broadcast Boris Yeltsin's address to the Army, encouraging it to resist the coup. Strangely enough, coup leaders did not shut down fax machines, jam the Voice of America broadcasts, or block telephone or computer communications. Information flowed freely to and from the rest of the world (Berkowitz and Quittner 1991). In this case, mass media joined the resisters in disseminating their messages.

Implications of Fax in China and Russia

Instant global communication through fax, short-wave radio, electronic mail, television, video, and telephone helped create the scene of the Chinese student movement in 1989. In Tiananmen Square, a pastiche was created of quotations from Mao, American songs such as "We Shall Overcome," and Lincoln's speeches; a statue, the Goddess of Democracy, was erected. At the beginning of the event, some primitive methods of disseminating information, such as wall posters, word-of-mouth, mail, and phone calls, were used. Later, short-wave radio, fax machines, video cameras, computer networks, and other new technologies were employed. Another interesting aspect of the flow of information in both the Tiananmen Square and the Soviet coup incidents was the mode of information. As Poster (1990) noted, the mode of information is recognized from decade to decade by variations in the structure of symbolic exchange:

> Every age employs forms of symbolic exchange which contain internal and external structures, means and relations of signification. Stages in the mode of information may be tentatively designated as follows: face-to-face, orally mediated exchange; written exchanges mediated by print; and electronically mediated exchange. (6)

The mode of transmission for Chinese dissidents was fax, video, TV, computers, and radio with all their postmodern signs and symbols. In a sense, all signs are information. Though fax can provide us with messages in graphics and different languages, its messages come in a structured textual form on paper, not that different from the paper and text used by Martin Luther to disseminate his Theses.

Another important aspect of the power of fax during the Tiananmen Square and the Soviet Coup incidents is how fax bypassed censorship. Satellite transmission was cut off in China before the massacre began. Though no images of the bloody incident could be transmitted then, fax was still working. The Chinese government failed to control the power of this technology. The Soviet coup leaders, likewise, did not bother to shut off fax machines, email, computer bulletin boards, and cable television (Belsie 1991). But, "you can't stop the flow of information by tanks" (Berkowitz and Quittner 1991). As de Sola Pool (1990) pointed out, "International communication is often considered a mixed blessing by rulers. Usually they want technical progress. They want computers. They want satellites. They want efficiently working telephones. They want television. But at the same time they do not want the ideas that come with them" (101).

References

Belsie, L. 1991. Technology thwarted coup leaders' success. The *Christian Science Monitor*, 26 August: 9.
Benjamin, D. 1986. State of siege. With Tiananmen Square the eipicenter, a political quake convulses China. *Time*, 29 May:36-45.
Benjamin, W. [1936] 1986. The work of art in the age of mechanical reproduction. In *Video culture*, edited by John G. Hanhardt. Layton, Utah: Peregrine Smith Books.
Berkowitz, H., and Quittner, J. 1991. Media and message. *Newsday*, 22 August: 14, 16.
Bowker annual book trade almanac. 1994. 39th ed. New Providence, NJ: R.R. Bowker.
de Sola Pool, I. 1983. *Technologies of freedom*. Cambridge, MA: Belknap Press of Harvard University Press.
———. Ed. 1977. *The social impact of the telephone*. Cambridge, MA: MIT Press.
———. 1990. *Technologies without boundaries: On telecommunications in a global age*. Cambridge, MA: Harvard University Press.

Martz, L. 1989. Revolution by information. *Newsweek*, 19 June 29.
Nichols, B. 1988. The work of culture in the age of cybernetic systems. *Screen*, 29 (1): 22-46.
Pacey, A. 1990. *Technology in world civilization: A thousand-year-history*. Cambridge, MA: MIT Press.
Pierce, J. R. 1977. The telephone and society in the past 100 years. In *The social impact of the telephone*, edited by Ithiel de Sola Pool. Cambridge, MA: MIT Press.
Poster, M. 1990. *The mode of information: Poststructuralism and social context*. Chicago: University of Chicago Press.
Toffler, A. 1990. *Powershift*. New York: Bantam Books.
———. 1980. *The third wave*. New York: William Morrow.
U.S. Bureau of the Census. 1991. *Statistical abstract of the United States*. 111th ed. Washington, DC.

Chapter Seven

Computer Links to the West: Experiences from Turkey

Marina Stock McIsaac
Petek Askar
Buket Akkoyunlu

> We must think of the whole of mankind as being a single body and of each nation as constituting a part of that body.
>
> —Kemal Ataturk

Computer-mediated communication (CMC) offers a tool, a technology for structuring social relations to provide a renewed sense of community. This community is a socially produced space in which people with similar backgrounds and interests construct narratives and dialogs in a highly mobile environment. CMC not only restructures human relations, it customizes social contacts using a postmodern geography. Electronic communities are reformulating traditional geographic communities. These new social formations of communities in cyberspace have a strong influence on people's forms of interaction. Women may develop stronger voices within an international community that supports their ideas. Academics in less prestigious disciplines may engage in discourse and dialog that can call into question previous notions of academic expertise.

CMC may also provide an environment for individuals to shape the community to which they choose to belong. Networks such as the Internet provide an environment for such a social construction of reality. Scholars are now examining electronic communities to determine the role that members take in such new communities and to explore the nature of their narratives. Cybercommunities can form a matrix of social relations that invites one to ask new questions about the nature of community. How can electronic, nontraditional social structures lead to a sense of global community? Who are we hoping to be as members of this global society? In this era of postmodern geographies, many researchers are examining the need for new communities and the technical possibility of creating them.

The formation of new communities in cyberspace provides unique opportunities to explore the role of developing countries such as Turkey in this socially produced space. Virtual mobility allows members of the electronic community to move from place to place as travelers, but also to move in status, class, and social position in an international society. This may be especially important as Turkey moves toward equality of educational opportunity for all its citizens.

The Turkish Experience

Turkey is an extraordinary country whose citizens have struggled for centuries to maintain their national identity. Even today they continue the struggle while surrounded by unstable political climates in neighboring countries: Iraq, Iran, Syria, and the former USSR.

The Western principles of democracy upon which Ataturk founded the Turkish Republic in 1923 included open communication and an education for all citizens to enable them to make informed decisions. Although Turkey has modeled its educational system after Europe and the United States, the country remains socially separated from the rest of the Western world. In fact, Turkey is still trying to become a full member in good standing of the European Union, but full membership depends on a number of factors, not the least of which is a well-educated populace that communicates with the outside world. Turkish education has been cited as one obstacle to full membership; foreign language proficiency is another.

Establishing the Internet

The Internet has increased communication and reduced isolationism for Turkey's academic and business communities since the 1980s. The growth of electronic networks is having profound effects on language, culture, and community by linking Turkey with communities around the world. It is influencing the nature of Turkish communication and giving voice to people who were previously unheard. For the first time, some citizens no longer need to travel abroad to discover certain aspects of other cultures. The Internet is helping to empower users with common interests, grouping them into virtual communities that cross geographic boundaries.

This chapter reports on the growth of the Internet in the Turkish academy and the manner in which email gave new voices to those academics with little social power by providing access to global communities.

Initial projects. Initial networking projects in Turkey began with the European Academic Research Network (EARN) node established in Izmir at Ege University during 1985. Early developments were tied to the progress of the telecommunications infrastructure and services (Tonta and Kurbanoglu 1995). Turkish Universities and Research Institution Network (TUVAKA) offered the first telecommunications services and was established to provide EARN links from Italy to Turkey. These services were offered to all universities and noncommercial research institutions without a fee. In 1986, the first connections were established to give Turkish universities access to networks such as Bitnet for communicating specifically with European and American academic and research institutions. The first group to join the network included twelve universities and TUBITAK, an umbrella research organization. These institutions quickly exhausted the resources of the available telecommunications provider.

University experience. In 1986, at the end of the first year of network connection in Turkey, Dr. Marina McIsaac conducted a project, "The Turkish University On-line," to examine how the Internet, as a new communication medium, affected various communities of users in a Turkish university. The project began with a Fulbright Fellowship granted to McIsaac in 1986-87 to teach and conduct research at a university in Turkey in a small city on the

Anatolian plateau. McIsaac designed a project to introduce the use of electronic mail to selected faculty and students in the Educational Communications Department. The project continued throughout a six-year period from 1986 to 1992 and highlighted the way in which networks were used, how use was allocated, who had access to computer networks, and who were the gatekeepers. The research asked whether new constituencies would be aided, how that would occur, and if the development of the Internet might provide opportunities for previously unheard voices. Initial data were collected by McIsaac, and follow-up information was provided by co-authors Petek Askar and Buket Akkoyunlu.

The results of observations and interviews indicated that the introduction and use of CMC by faculty and students in educational communication at the university resulted in new learning opportunities for those previously disenfranchised faculty and students who had not had access to the technology. At Turkish universities, faculty in the social sciences are traditionally considered to have less use for computers than their physical scientist counterparts. Social scientists have not been provided with access to computers to the same extent as their colleagues in medicine, engineering, and physics. In this project, the use of electronic mail helped to liberate social scientists by mitigating the status differential between them and physical scientists. This was evident in the patterns of computer use generated by the social science faculty. The gap which was previously felt in the social sciences began to narrow as faculty in the Educational Communications Department participated in their own academic networks.

The absence of censorship fostered another area of equality. In Turkey at that time it was a punishable offense to criticize the government, so radio and television stations were highly regulated. Although electronic mail was never totally free from observation, it remained a more open environment for communication than published or broadcast media.

Women at the university, both faculty and students, felt the egalitarian effects of email because, in their opinion, it was a technology that helped to eliminate gender bias. Some women who felt intimidated in face-to-face settings were able to exchange views by signing messages with their initials, rather than using their female names.

Women received support for their views from colleagues at other universities and felt they could engage in more intellectual conversations in which their opinions were taken seriously. In addition, they noted support and affirmation of their views from international peers.

Both male and female students in the social sciences described feeling liberated both intellectually and culturally. They experienced freedom from intellectual constraints imposed by an academy that, in some cases, had narrowly defined areas of scholarship in terms of physical sciences. These students were able to rely on outside scholars to provide a wider knowledge base and to stimulate analytic thinking. In addition, the use of CMC allowed a large number of students to free themselves from cultural restraints. In a Muslim country like Turkey, topics such as sex are not considered appropriate for general discussion. Even dating is not encouraged in rural communities and same-sex friendships provide the bulk of social activities both before and after marriage. There is a lack of adequate information about sex, so students sought that information on the network. Students who used CMC quickly found the sex-related information they were seeking and discovered that they could ask questions in a nonthreatening atmosphere without the cultural restraints they experienced in their own environment.

New alliances. At the beginning of the CMC project with faculty in the Educational Communications Department, the most enthusiastic response came from those who had studied abroad and were eager to re-establish links with their foreign colleagues. The next most enthusiastic group were faculty and graduate teaching assistants who planned to study abroad and wanted to establish international contacts. There was a heavy demand for articles and published materials from Europe, the United Kingdom, and the United States. In Turkey, the average cost of a textbook published abroad could be as much as one-half of an assistant professor's monthly salary. Those who studied in the United Kingdom or United States returned with reference materials, but were subsequently unable to remain current in their disciplines. International coalitions were seen as the only opportunity to keep up to date with new research.

A second benefit of such coalitions among faculty was the gradual development of their voices in foreign academic publishing communities. Through liaisons formed with internationally respected peers, some faculty were able to publish abroad (in Turkish academic circles, publishing in British or U.S. journals is of utmost importance for academic recognition). This achievement required frequent international communication, which was almost impossible considering that round-trip airmail took one month or more. CMC facilitated this communication, since it was fast and informal.

Women. Women who participated in this project reported the delight of making friends abroad. In a number of cases, these women were able to obtain information from foreign colleagues more readily than from colleagues at their own institution. In addition, they received invitations to participate in international meetings and seminars. Because of gender biases at home, these women did not receive such invitations from their Turkish colleagues. Women talked in particular about the importance of networking with other women worldwide.

Students. Students reported the advantages of using CMC for practicing language skills, gaining exposure to international standards for academic study, and developing a greater understanding of others by sharing divergent worldviews. On the other hand, a number of students reported that they felt the competence they gained in the international arena from their use of electronic mail was often viewed as threatening by their professors who did not use CMC.

Although the initial purpose of the project was to introduce a new technology, the subsequent utilization of computer-mediated communication (CMC) had social, political, and cultural implications far beyond the university context. By the end of the project, faculty and students had developed academic ties and social relations around the globe. They were able to expand their academic knowledge and develop online collaborations that gave them a broader view of the world. New users described the advantages of CMC as expanding their horizons by providing local online resources as well as colleagues in other countries with whom they could form new national and international coalitions.

Recent Growth and Directions

The initial project, "Turkish University On-line," was completed in 1992. During that initial six-year period, changes occurred very slowly. Since 1992 there has been substantial growth in services and the number of Internet users. As networks become overcrowded, new services have emerged to provide additional telecommunications access. In 1996, it was estimated that in Turkey 13,000 computers had access to the Internet with 100,000 users, at over 800 sites (Ozdogan 1996). Since 1995, the largest growth in users has been in the commercial sector. There are a number of significant trends in the growth of the Internet.

Computer Experimental Schools Project (CES). The CES project is one of a number of projects being implemented as part of the Turkish Government's National Education Development Project under a loan agreement with the World Bank. The goal of the CES project is to provide schools with computers and modems so that all schools will have equal access to information and educational materials (Askar, Rehbein, and Noel 1996). Rural schools and those in remote areas will be given priority as funds become available. A wide area network (WAN) will be piloted for experimental computer schools as soon as possible.

Use of technology for distance education. The demographics of computer and network use in Turkey will change as new telecommunications networks spread. A recent survey of 2000 students from private, public, elementary, secondary, and vocational schools in both rural and urban Ankara (the capital of Turkey) revealed that 47% of the students are familiar with computers, but only half of them have a computer in their homes. In private schools, 83% of the students use computers. In Anatolian high (semiprivate) schools 69% of students have access to computers, and in public schools only 22%. Of the students who use computers in schools, 56% have parents who have a university education; 21% have parents with a high school education; 9% have parents with a middle school education; 11% have parents with an elementary education; and .2% of the parents have no education. Of the girls, 45% use computers and 55% do not. Of the boys, 53% use computers and 47% do not. Only 2.5% of the students use the Internet for finding resources for their school work. When asked what the Internet is, 78% of students responded that they know that it is communicating via computers. Of teachers, 94% responded that they have heard about the Internet, but have no idea how they might use it in their teaching (Askar, Rehbein, and Noel 1996).

It is true that the use of computers is currently limited to a small group of students who are of high socioeconomic status and have parents with high educational levels. The number of Internet users is even smaller. However, there is a great deal of pressure on the Ministry of Education to provide computers and Internet access to schools. Pressure is coming not only from within Turkey but from outside as well, as Turkey tries to compete economically in the world market. Turkey, with a loan from the World Bank, has invested in an ambitious program of educational reform and telecommunications services involving massive purchases of computers and training programs for teachers. Included in these plans are initiatives for distance education programs involving telecommunication technologies that will offer education to both school age youngsters and adults. In addition to the university and high school distance education programs, adult continuing education will be available by distance learning technologies as well (Ozar and Askar 1997).

Foreign Languages

Turkish secondary education does not currently require that graduates be proficient in foreign languages, although foreign language courses are always required in the curriculum. This lack has been a barrier in communicating with people outside of the country. The Turkish language is not widely spoken outside Turkey. Turkish "guestworkers" who have left Turkey to earn a living in Germany have brought the German language back home, and there is a historical tradition of speaking both English and French. However, until recently, international communication was limited to businessmen, guestworkers abroad, some academics who traveled overseas, and the wealthy.

Since the mid 1980s there has been increasing emphasis on learning foreign languages, particularly English. Since 1980 the tourist industry has grown more than 100%. Turkish tourism workers now need both German and English to be successful. In addition to those in the tourist industry, students, businesspeople, and professionals have traveled abroad to the United Kingdom, the United States, Asia, and Europe. These Turks have expanded their horizons, learned foreign languages, and been introduced to new values, expectations, and lifestyles. Communication with people in other parts of the world has become increasingly important to combat the isolationism that Turkey once felt.

Women and Cyberspace

Although still uncommon, the number of studies about social and cultural issues related to the use of electronic networks by previously disenfranchised users is increasing. In a recent unpublished master's thesis, "Gender and Cyberspace: A study on the participation of women in computer mediated communication," Duygu Alparslan (1997), explored issues related to women and their voices in cyberspace. Kadin, a women's discussion listserv in the Turkish language, was formed in August 1995 to "provide a democratic information and (a) discussion medium for people who are interested in and study 'women in Turkey' and 'women's problems.'"

By June 1996, 121 members had subscribed to the list. Interestingly, 51% of the members on the list were men, although the topics discussed were those concerning women (the gender of 7% of the subscribers was undetermined). There was no mistaking the purpose of the listserv, since it was called kadin (women). Only 42% of the subscribers to the listserv were identified as women, and they sent only 28% of the messages, whereas men sent 72% of the messages. This suggests that CMC as used in Turkey is not a gender-free medium of communication, but that it follows traditional Turkish cultural patterns. Women are less likely to speak out in a mixed forum where men are present. Men take the initiative to voice their opinions even though, as in this case, this was a women's listserv. Women generally remain less vocal, although there is a strong movement among women in professional positions to develop stronger women's voices among their colleagues. Although men communicated more than twice as often as women on this listserv, there was no evidence of any type of sexual harassment, perhaps because everyone used an institutional account where they were required to use their real names, so the identity of the correspondent was known to all.

An informal questionnaire. A questionnaire was sent to a small sample of these men and women CMC users. Three questions were asked:

- Do you think cyberspace is a genderless medium?

- Is communicating with the opposite sex in cyberspace different than in other mediums?

- Do you think that cyberspace is a male-dominated medium?

Responses to the first question indicated that most of the women thought that cyberspace is a gendered medium, that it has been and continues to be dominated by men, and that people tend to model the same gendered roles that they have in traditional communication. They are socialized to behave the same way, whether they are in cyberspace or face-to-face. According to the respondents in this survey, women follow the same traditions online that they do face-to-face. Women apologize more, are more tolerant, and are not dominant while men are more aggressive, dominate conversations, and express their views more freely.

Most respondents to the second question agreed that communicating in cyberspace is very different from traditional communication because of the lack of visual cues. Some women said that men appear to make sexual references more frequently over e-mail. "Many men find courage to sexually express themselves, otherwise they cannot manage to do it in real life" (Alparslan 1997, 24). Men agreed that communication was different in cyberspace and they mentioned that digital love affairs are more frequent than real life love affairs "maybe because cyberworld is a dreamland" (25).

In the third question, the majority of both men and women agreed that cyberspace is male dominated, since women don't participate as frequently. They also noted that more women are "lurkers" and do not voice their opinions as often as men in a mixed gender listserv. Perhaps it is only through interactions outside the existing social structure, where there is no fear of reprisal, that freedom of expression can exist. The larger international cyberspace community could provide that arena and encourage building communities based on common interests, intellectual collaboration, and shared resources.

Access

The "Gender and Cyberspace" study conducted in September 1995 suggested that in Turkey cyberspace is a gendered medium that perpetuates existing male and female roles and reflects the ways that those gender roles are socialized in society. It appeared that within the country gendered roles exist in computer-mediated communication as they do in traditional communication.

In the two years since the data were collected, CMC has become available to many more people, and use has grown for both men and women. There are now almost fifty Internet domains belonging to Turkish educational institutions. Some twenty governmental bodies, as well as banks, ministries, and public agencies are now on the Internet. Although demand for and use of the Internet are growing rapidly, users of the Internet generally have academic, professional, or governmental credentials. Economic constraints continue to slow the adoption of this form of communication. In general, women and other disenfranchised groups still have less access to the Internet than men, although this is slowly changing.

Education and Equality

Turkey is facing challenges on many fronts. Demographic studies show an increased migration of the poor to cities, high inflation rates, and an uncertain economy. With a consumer price index that rose 77% and interest rates that rose 108% between 1997 and 1998, there is a particular need to examine how education might help stabilize the economy. What form of training for the unemployed is most likely to be useful? In particular, educators are looking at ways to develop sound educational programs that will prepare workers for jobs in the future.

One measure of the health of a country's economy is the relationship between education and jobs. Turkish education has long been criticized for not providing a foundation for the transition to work. Of the twenty-five member countries of the OECD (Organization for Economic Cooperation and Development), Turkey had the lowest rate of sixteen-year-olds in school in 1992, the latest year that these figures were published. Only 39% of Turkish sixteen-year-olds were enrolled in secondary schools. Of the twenty other countries reporting, the percent of enrolled sixteen-year-olds ranged from 75 to 97 (OECD Observer, 1996). Young people who leave school at such an early age are often left without knowledge or skills to get useful jobs (McIsaac 1996).

In order to improve economic performance and competitiveness in international markets, educators and government officials must look to the emergence of a growing knowledge-based society. It is apparent that there is a current worldwide trend to enhance service sector jobs by emphasizing the value of technology and the importance of human resources. Jean-Claude Paye, Secretary-General of the OECD, notes that the business and telecommunications service industries, which are the primary purchasers of information and communication technologies, now account for more than a quarter of all monies spent in OECD countries. The world is fast becoming a knowledge-based economy. With the vast percentage of expenditures going into telecommunications and related information technologies, countries that want to be competitive must enhance the skills of their labor force. This goal can be accomplished by providing more efficient and equitable education, training, and skills development. In this way, people can be prepared for jobs using information technologies as new products and services are created. Lifelong learning will be required as skills become obsolete and are replaced by new products and services. Technology-based communication is attractive to developing countries and offers a way to equalize access to information in a cost effective manner. Studies like the "Turkish University On-line" and "Gender and Cyberspace" add to an understanding of how the Internet can be used to empower currently underserved groups, and how it may give them the tools to expand their knowledge and communicate with their peers around the world.

Democratization. Although electronic networks were established primarily for engineers, doctors, and physical scientists, it is clear that the technology has different benefits for those with lower academic status. Beyond providing academic information, the use of CMC provides the opportunity for addressing equity issues, offers an equitable channel for social information, and broadens cultural perspectives. The extent to which these social elements of telecommunications technology continue to provide opportunities for democratization will depend on access to the technology. Turkey is a country of contradictions, with one foot firmly planted in Asia and the other tentatively moving toward Europe. Telecommunications offers the potential to promote the country's socioeconomic goals and democratic ideals.

References

Alparslan, D. 1997. Gender and cyberspace: A study on the participation of women in computer mediated communication. Unpublished Master's Thesis. Middle East Technical University, Ankara, Turkey.

Askar, P., Rehbein, L., and Noel, K. 1996. CES Project: Mid-term review and evaluation consultants' report. Ankara, Turkey: World Bank.

McIsaac, M. S. 1996. Learning for the future. Keynote address at the First International Distance Education Conference, Ankara, Turkey, November. (http://seamonkey.ed.asu.edu/~mcisaac/futures).

———. 1993. Economic, political and social considerations in the use of global computer-based distance education. In *Computers in education: Social, political, historical perspectives,* edited by R. Mufolleto and N. Knupfer. Cresskill, NJ: Hampton Press.

OECD. 1996. Economic survey of Turkey (an OECD economic survey). Located at http://www-oecd/search9cgi/s96.

Ozar, M., and Askar, P. In press. Present and future prospects of the use of IT in schools in Turkey. International Review section, *Educational Technology: Research and Development: A Quarterly Publication of the Association for Educational Communications and Technology.*

Ozdogan, H. 1996. Seminar on the Internet. Ankara, Turkey, December.

Tonta, Y., and Kurbanoglu, S. 1995. Networked information in Turkey. *Journal of Turkish Librarianship:* 230-234.

Contributors

Dr. Ann De Vaney is Professor Emerita in the Department of Curriculum and Instruction at University of Wisconsin, Madison and Visiting Professor Emerita in the Department of Education at the University of California, Irvine. Her work explores the intersections of education, technology, and social/cultural issues. More specifically she employs discursive and rhetorical techniques to explore issues of representation in educational and popular media and technology texts. She is the author of numerous articles in scholarly journals and chapters in books, and has edited academic books and journals. Her most recent publications include an article in *Educational Policy* (Sept. 1998) and a guest-edited issue of *Theory into Practice, Technology and the Culture of Classrooms* (Winter, 1998).

Stephen Gance received his B.S. in mathematics and M.S. in computer science from the University of Colorado in Boulder. He is currently a doctoral student in the Educational Communications and Technology area in the Department of Curriculum and Instruction at the University of Wisconsin-Madison. His dissertation explores educational technology texts to locate how discourses work within these texts to construct particular visions of the technological subjectivities of teachers and students in a pedagogical relation to educational technology.

Sousan Arafeh received her B.A. from Hampshire College and her M.A. from the University of British Columbia (UBC). She is currently a doctoral student studying policy making, telecommunications, cultural studies, new media, and education in the Department of Communication Arts and the Department of Curriculum and Instruction at the University of Wisconsin-Madison (UW-M). She is a UW-M Avril S. Barr Fellow and was awarded a University Graduate Fellowship at UBC. Arafeh has presented numerous papers and is publishing articles on aspects of telecommunications policy and education. She also chaired the city of Madison, Wisconsin's Broadband Telecommunications Regulatory Board.

Dr. Kedmon Hungwe is a senior lecturer and chairperson at the Center for Educational Technology, University of Zimbabwe. He did his initial graduate work at the University of Wisconsin-Madison, and later at Michigan State University where he completed a PhD in Educational Psychology. His research has focused on the relation between education, media, and social change. He has published in several journals.

Dr. Stephen T. Kerr is Professor of Education at the University of Washington, from which he also received his PhD. His research focuses on how technological changes affect the social structure of educational institutions. His work has appeared in many academic journals, and he was contributing editor for *Technology and the Future of Schooling*, 95th Yearbook of the NSSE. He has worked extensively in Russia since 1978, focusing there on how mechanisms of educational renewal emerge under difficult material and intellectual conditions.

Dr. Marina Stock McIsaac is Professor of Educational Media and Computers at Arizona State University where she teaches and conducts research on computer applications in education, telecommunications, and distance education. She has been the recipient of two Fulbright Senior Scholar awards to lecture and conduct research on distance education and educational technology in Turkey. Dr. McIsaac has lectured in Italy, Germany, France, Turkey, and Australia, serving as a consultant for the National Institute for World Trade, the Turkish Ministry of Education, the USIA, and the United States Department of Education. McIsaac has published in numerous scholarly journals and is past president of the International Division of the Association for Educational Communication and Technology, the Research and Theory Division of the Association for Educational Communications and Technology, and the International Visual Literacy Association.

Dr. Petek Askar is a Professor in the Department of Science Education, Faculty of Education at Middle East Technical University in Ankara, Turkey; she has also served as a consultant to the Turkish Ministry of Education in their "Computers in the Schools" initiative. She has conducted research and published widely on computers and education.

Dr. Buket Akkoyunlu is Associate Professor of Computer Education and Instructional Technology at Haceteppe University in Ankara, Turkey. She has published in the areas of information technologies, using computers in schools, and educational technology in Turkey for primary school teachers.

Dr. Yan Ma is Associate Professor in the Graduate School of Library and Information Studies at the University of Rhode Island. She has a BA in English Literature and Linguistics from ZhejiangUnivesity, an MLS in Library and Information Science from Kent State University, and a PhD in Curriculum and Instruction from University of Wisconsin, Madison. She has brought her skills in structural and poststructural analysis to the areas of library and science and educational technology to address issues of culture. She has also published widely in academic journals.

Zarni, a Bumese dissident-in-exile, founded the Free Burma Coalition. He holds a B.Sc. in Chemistry from Mandalay University and an M.A. in Education from the University of California at Davis. Recently he earned a Ph.D. in Curriculum and Instruction from the University of Wisconsin-Madison, having written a dissertation titled "Knowledge, Control and Power: The Politics of Education under Burma's Military Dictatorship (1962-88)."

Index

A

Abramov, Alexander, 105
Academy of Pedagogical Sciences of the USSR, 92, 103, 104
Adamsky, Alexander, 96, 97, 102
    Adamsky's Eureka group, 100, 101–102
Adorno, Theodor, 28
Aesopian lanuage, 91
Africa Watch, 52, 58, 63
African National Congress (ANC), 17
Akkoyunlu, Buket, 153, 156
All Burma Students Democratic Front (ABSDF), 73
    information collection and dissemination, 75–76
Alliance for Democracy (AFORD), 61
Alparslan, Duygu, 161, 162
Appaduri, Arjun, 11
Arafeh, Sousan, 113, 167
Askar, P., Rehbein, L., and Noel, K., 159, 160
Askar, Petek, 153, 156, 168–169
Ataturk, Kemal, 153, 154
Aung San Suu Kyi, 76, 81, 84
    cofounder (NLD), 74
    endorsement of fast, 85
    Nobel Peace Prize, 75
    release from house arrest, 79

B

Banda, Hastings Kamuzu, 51, 52, 56
    allies, 57
    beginning of repressive rule, 56–57
    censorship and control, 58
    education, 53
    emergence of, 53
    end of rule, 67
    return to Malawi, 53, 55
Barron, S., 71
Baumeister, Rolf F., 32
Bell, Alexander Graham, 131
Belsie, L., 150
Benhabib, Seyla, 38
Benjamin, D., 140, 141
    about the Tiananmen demonstration, 139
Benjamin, Walter, 144, 145
Berkowitz, H., and Quittner, J., 148, 149, 150
Bhabha, H., 132
Bimber, Bruce, 26–27
Burma, 71–86
    1989 and rising political repression, 74–75
    independence of, 72
    "State Law and Order Restoration Council" (SLORC), 74
    suppression of dissent, 73
Burma-focused nongovernmental organizations (NGOs)
    aid to Burmese refugees, 73
    information collection and dissemination, 75–76
Burma Socialist Program Party (BSPP), 72
BurmaNet, 75–76, 77–78
BurmaNet Conference, 77
Burmese exiles. *See also* Free Burma Movement

All Burma Students Democratic Front (ABSDF), 73–74
  beginnings of an international movement, 73–74
  Committee for the Restoration of Democracy in Burma (CRDB), 73–74
  EuroBurmaNet, 85
  formation of email newsgroup, 75
  Interent information dissemination, 75–77
  Internet use, 71–72, 75–86
Butler, J., 32, 128, 131
Bwanausi, Augustine, 56

C
Carlyle, Thomas, 23–26, 28, 29
  influences of, 24
  predecessors to, 24–25
Castells, Manuel, 11, 19
  civil rights movements, 40
  on identity, 34, 38–39
  Internet host in South Africa, 18
  on subjectivity, 32, 34–35
  success of groups and specific rights, 43
  vision of resisters, 35
Center for Democracy Washington, 148
Charles, D., 71
Chihana, Chakufwa, 61–62, 64
  arrest of, 62
China
  invention of movable type printing, 143
  invention of paper, 143
China and political dissent, 139–150
  Chinese dissidents pursuit of democracy and online presence, 139–141
  communication technologies and citizens, 142–149
  fax and American Chinese-language newspapers, 146–148
  fax invisibility to censors during Tiannanmen, 150
  implications of fax, 149–150
  Mertz's technologies of freedom, 146
  Tiananmen group identity, 34
  Tiananmen Square and the use of fax technology, 141–142, 146–148
  use of phones, 143
Chipembere, Henry, 56
Chipembere, Masauko, 55, 56
Chipeta, Mapopa, 64
Chirunga Newsleter, 58–59
Chisiza, Dunduzu, 56
Chituko Cha Amai Mu Malawi (CCAM), 63
civil rights movements, 40
Claisse and Rowe
  instrumental phone use, 123
Cleaver, 18–19
Committee for the Restoration of Democracy in Burma (CRDB), 73
communication technologies and citizens, 142–149. *See also* China and political dissent
  assessing decentralized communication technologies, 143–145
  de Sola Pool, 143–144, 145

decentralized technologies,
142–143
decentralized technology and
resistance, 145–146
implications of fax in China
and Russia, 149–150
mode of information, 149
Russia and the use of fax
technology, 148–149
"The Art of Reproduction in
the Mechanical Age"
(Benjamin, W.), 144
"The Work of Culture in the
Age of Cybernetic
Systems" (Nichols), 144
computer-mediated communication (CMC), 153–154,
158, 161
equity issues, 165
freedom from cultural restraints, 157
"Gender and Cyberspace"
(Alparslan), 161, 164
introduction and use by
faculty and students,
156
new availability of, 163
questionnaire for users, 162
Cooley, 32
Cybercommunities, 77–85, 154
Cybersex: An adult affair, 129

D
Daily Times, 58
Davydov, Vasily, 99
de Certeau, M., 133–134
de Condorcet, 24–25, 26
de Sola Pool, I., 118, 143, 144,
150
decentralized technologies
assessment of, 143–145

books, 143
computers, 144
resistance in Eastern Europe
and China, 145–146
telephones, 143–144
democracy
a diffuse modern sign, 15–16
democracy and equality
in new resistance movements, 41–44
democracy and technology
discourses and diversity, 37–
38
influences of Enlightment
rhetoric, 12–15
modern deterministic
threads, 14
modern resistence movements, 29–30, 40
democratic politics
resurgence of interest, 12
DeVaney, Ann, 9, 31, 167
Dimmick, J., Sikand, J., and
Patterson, S.
intrinsic phone use, 123
"Discourse on the Progress of
the Human Spirit"
(Turgot), 24
Dneprov, Eduard
historian of Russian education, 104
Drummond, E., 59

E
Eckholm, Eric, 21, 22, 38
Edwards, W., 126
El'konin, 99
Ellsworth, Elizabeth, 31
Ellul, Jacques, 24, 28, 29, 36
equality and democracy

in new resistance movements, 41–44
"Eureka" (*Evrika*)
  Russian educational reform, 97
EuroBurmaNet, 85

F
fax technology
  Malawi, 51, 63–68
  transition to newspapers in Malawi, 67–68
Fedorov, Ivan
  Russia's first printer, 90
Ferguson, Robert A., 14
Fish, Stanley, 31
Fisher, C.S., 118, 119
Franklin, Benjamin
  democracy and technology, 12–14
Fraser, N.
  modifications to public-sphere theory, 117–118
Free Burma Coalition (FBC), 83, 84, 85
  creation and establishment of, 79–81
  motto, 84–85
  purpose of, 79
"Free Burma Community," 85
Free Burma Movement
  Chapel Hill demonstration, 81
  consumer boycotts, 77–78
  creating an identity, 84–85
  developing a presence on the Net, 81
  diversity in the movement, 83–84
  EuroBurmaNet, 85
  Free Burma listserv, 83
  growth of communities in cyberspace, 77–79
  "International Day of Action for a Free Burma," 80
  need for a centralizing body, 78–79
  non-Burmese activists in Burma, 76
  resistance and cybercommunities, 71–86
  Student Environmental Action Coalition (SEAC) conference, 80–81
  Web site for Action Day, 81, 83

G
Gance, Stephen, 1, 167
"Gender and Cyberspace: A study on the participation of women in computer mediated communication (Alparslan), 161, 164
George Soros's Open Society Institute
  Russian educational reform, 105, 106
Gill, M.S., 128
Girshovich, Vladimir, 105
Gorbachev, Mikhail, 93, 103, 104, 148
Government of National Unity (GNU), 18
Gromyko, Yuri, 100

H
Habermas, Jurgen, 29, 116
Hall, Mike, 62–63
Hamilton, Alexander

democracy and technology, 13–14
Hanson, E., 129–130
Haraway, D., 128, 131, 132
Havel, Vaclav, 85
Heidegger, Martin, 29
Heller, K., Thompson, M., and Trueba, P., 125
Herrnstein Smith, Barbara, 12
Horkheimer, Max, 28
Horkheimer, Max, and Adorno, Theodor, 24, 29
Hungwe, Kedmon N., 16, 51, 168

I
Iagodin, Gennady, 97
identity, 32, 34, 35, 128
   telephone/telephone user and, 130–131
information technologies, 141, 144–146, 149–150
   centralized state bureau, 11
   employment and, 164
   fax, 17, 51, 62–68, 141–142, 146–150
   Internet, 71–86, 89–90, 140–141, 155–165
   mode of information, 149
   new discursive space and multiple subjects, 38–39
   print media, 58–63, 67–68, 89–108, 141–143
   resistance movements, 11
   telephone, 113–134, 141, 143
Internet
   Free Burma Movement, use of, 71–72, 75–86
   overblown claims about, 10–11
Iser, Wolfgang, 31

J
James, William, 32
Jauss, Hans, 31
Jefferson, Thomas, 140
   democracy and technology, 12–14

K
Kadzamira, Cecilia, 58, 63
Kamkondo, Dede, 65
Kant, 116
Kanyama, Chiume, 55, 56–57
Katz, J.E., 127
Kerr, Stephen T., 89, 168
Kinelev, V.G., 106

L
Latour, B., 128
Lawyers Committee for Human Rights, 58
Lincoln, Abraham, 34, 140
"Living Our Faith"
   pastoral letter banned, 60
   trigger for political change in Malawi, 59–61
Long, Simone, 85
Luria, 99, 100
Lusaka, 61–62, 63
Luther, Martin, 149
Lwanda, J.L., 51, 58, 59
   Bishop's letter, 61
   government raid of companies, 67

M
Ma, Yan, 139, 169
Malawi, 16, 51–68
   background politics, 53
   Banda's return, 53, 55

beginning of Banda's repressive rule, 56–57
Catholic church's protest against government, 59–61
censorship and control under Banda, 58
coalescing of colonial opposition, 55–56
colonial politics, 53
end of Banda' rule, 67
fax technology and opposition, 63–68
geographic description, 52
influences of changes in Eastern Europe, 52, 58
Malawi Censorship Board, 58
nongovernment sponsored publications, 51, 58–61, 63–67
political transition, 67–68
resurgence of organized political opposition, 61–62
resurgence of political resistance, 58–61
state-controlled newspapers, 58
state-controlled radio station, 58
Malawi Broadcasting Corporation, 58
Malawi Censorship Board, 58
Malawi Congress Party (MCP), 52, 61
end of, 67
Malawi Democrat
an opposition newspaper, 63–64
Frank Mayinga and Mapopa Chipeta, 64

Malawi News, 58
Malawi Young Pioneers (MYP), 57
Mandela, Nelson, 44, 58
Mapanje, Jack, 65
Marcuse, Herbert, 24, 28, 29
markets and democratic political discourses, 20–22
China, 21–22
Martin, M., 118, 121, 122, 128
men and phones, 119
romance-related calls, 126
Martz, L., 142
technologies of freedom, 146
Marx, Karl, 26–28
Marx, Leo, 23, 27–28
Marx, Leo, and Smith, Michael, 13
Marxism, 20
Matveev, Vladimir Fedorovich, 95, 97, 98, 99
Mayinga, Frank, 64
McIsaac, M.S., 153, 155–156, 164, 168
McLuhan, 142, 143
Mead, George Herbert, 32
Mhango, Alkwapatira, 57
MIROS (the Moscow Institute for the Development of Educational Systems)
Russian educational reform, 105
Morley, D., 116
Moyal, A., 118, 122, 128
study of Australian women and phone use, 124–125
Mpakati, Attati, 57
Mumford, Lewis, 29

N
Namibia Today, 17

National League for Democracy (NLD), 74, 76
National Unity Party, 74
Newell, J., 59
Nichols, Bill, 144, 145
Nobel, G., 123
nongovernmental organizations (NGOs), 18–20, 41
  aid to Burmese refugees, 73
Novaia Shkola (New School) publications, 105
  Russian educational reform, 105

O
O'Keefe, G., and Sulanowski, B.
  intrinsic phone use, 123
Open Society Institute, 75
Organization for Economic Cooperation and Development (OECD), 163–164
Ozar, M., and Askar, P., 160
Ozdogan, H., 159

P
Pacey, A., 143
Paye, Jean-Claude, 164
Peck, John, 81
Pedagogical Kaleidoscope, 105
Pedagogicheskii vestnik, 105
"Pedagogy of Cooperation"
  russian educational reform, 96
Peikan, Jiang, 38
PepsiCo Inc., 79, 80
perestroika, 93
Pervoe sentiabria (September First)
  Russian educational reform, 98–99, 105
political candidates

promotional Web sites, 11
Polozhevets, Petr, 98
Porteus, Kim, 17–18
Poster, M
  on mode of information, 149
Postman, Neil, 24
postmodern world
  aims of postmodern scholars, 12
  resistance movements, 11, 30
Pratt, Carrie, 17–18
Price, Todd, 80
print media
  value of, 90

R
race
  democratic discourses, 37–39
Rakow, L., 118, 128
  early phone company campaigns, 119
  quotation about women's engagement with telephone, 120, 121–122
resistance movements
  agency of resistors, 36–37
  heir to wisdom of civil right movements, 40–41
  information technology, 11
  interpretations of democracy and equality, 41–44
  postmodern world, 11, 30
  subjects in, 35–36
Richards, A.K., 128
Rixaca, 17
Roche, J., Monsignor, 61
Ronell, A., 130, 131
Rorty, Richard, 38
Rubin, R., Perse, E., and Barbato, C., 123
Rubtsov, Vitaly, 100

Russia, 89–108
  Academy of Pedagogical Sciences of the USSR, 92, 103
  dynamics created from censorship, 91–92
  early cultural forms, 91
  educational challenges to central control, 93
  the fate of print in an electronic world, 107–108
  fax by-passed by censors during coup attempt, 150
  implications of fax, 149–150
  perestroika, 93
  role of print media, 90–91, 107–108
  "Socialist Realism," 92
  state control and education, 92–93
  tensions between printed texts and politics, 91–93
  "Theory of Conflictlessness" (*Teoriia bezkonfliktnosti*), 92
  use of Aesopian language, 91
  use of fax during 1991 coup attempt, 148–150
  value of print media, 90
  "wall newspaper," Soviet propaganda device, 102
Russian educational reform, 94–106
  an unsettled future, 102–106
  beginnings and causes of, 94–95
  Eduard Dneprov, 104
  emergence of alternative publication outlets, 105–106
  "Eureka" (*Evrika*), 97
  George Soros's Open Society Institute, 105, 106
  importance of textbooks, 106
  influences within academic circles, 99–100
  lack of infrastructure and capital, 104–105
  MIROS (the Moscow Institute for the Development of Educational Systems), 105
  need for new textbooks, 103–106
  new programs for teacher education, 100
  new teacher programs foundations in print traditions, 102–103
  Novaia Shkola (New School), 105
  "organizational-activity games" (*organizatsionno-deiatel'nostnye igry*), 101
  Pedagogical Kaleidoscope, 105
  "Pedagogy of Cooperation," 96
  Pervoe sentiabria (September First), 98–99, 105
  print and the rise of activity theory, 100–102
  Psychological Institute of the Academy of Pedagogical Sciences, 99
  rise of teacher professionalism, 97
  "the movement" (*obshchestvenno-pedagogicheskoe dvizhenie*), 96

"The Union of Teachers" (The Creative Union of Teachers), 97–98
Uchitel' skaia gazeta (Teachers' Gazette), 95–96, 98–99
Vasily Davydov, 99
VNIK-Shkola and curriculum reform, 103–104
World of Education, 105

S
Sarch, A., 118, 126
Schmitt, 29
Scott, D., 125
Seleznev, Gennady, 98
Sen, Amartya, 42
September First (Pervoe sentiabria)
  Russian educational reform, 98–99
Seregny, Scott, 93
Shchedrovitsky, G.P., 100
Shefield, C., 127
Sheng, Bi, inventor of movable type printing143
"Signs of the Times" (Carlyle), 23–24
"Sketch of a Historical Drama of the Progress of the Human Spirit" (de Condorcet), 24
Smith, M., and Morra, N., 127
Smith, Michael, 13
Soloveichik, Simon, 95, 97, 98
South Africa and Namibia, 16–18
  democratic goals, 18
South West African People's Organization (SWAPO), 17
Soviet Union. *See* Russia

Spender, 116
"State Law and Order Restoration Council" (SLORC)
  Rangoon, 74
Steele, Douglas
  establishment of BurmaNet, 75
Stone, A.R., 129, 130, 131
Student Environmental Action Coalition (SEAC)
  Free Burma Movement, 80–81
subjectivity and identity, 30–44, 128
  Castells on, 32, 34–35
  identity, 32, 34, 35
  race, 37–39
  subjects, 31–32
  symbolic interactionism, 32–33
  telephone technology and, 130–132
  Touraine on, 32–34, 35
  subjects, 31–32
  resistance movements, 35–36
Sukhomlinsky, 100
symbolic interactionism, 32–33

T
Teachers' Gazette (Uchitel' skaia gazeta), 95–96, 98–99
Technica, 17
technology and revolution
  Marx, Karl, 26–27
technology discourses, 22–30, 35–37
  Carlyle, 23–25
  determinism, 24
  modern, 22, 25–30
  resistence movements, 29–30

technology and revolution, 26–27
Weber, 28–29
"Technology in Old Democratic Discourses and Current Resistance Narratives"
  categories of resistance movements, 16
  purpose of, 9–10, 12
telephone technology, 113–134
  as a communicational tool, 115
  definitions of public and private space, 116–118
  expanding subjectivity, 131–132
  as a gendered technology, 116
  men's relationship with, 119, 120, 123, 126
  phone and indentity, 130–131
  a postmodern approach, 130–131
  relationship to Internet, 115
  representations of, 115–116
  solidarities and the formation of quasi-public spheres, 132–134
  telephone penetration - democraphics, 114–115
  use as "quasi-public" space, 113, 118, 125, 126, 127
  uses by gay men, 129–130
telephone technology and women. *See* women and telephone technology
Tembo, John, 63
Teoriia bezkonfliktnosti ("Theory of Conflictlessness), 92

"The Art of Reproduction in the Mechanical Age" (Benjamin, W.), 144
The Economist, 85
The Post
  independent Zambian paper, 63
"The Union of Teachers" (The Creative Union of Teachers)
  Russian educational reform, 97–98
The Vacant Seat (Kamkondo), 65
'The Work of Culture in the Age of Cybernetic Systems" (Nichols), 144
Tiananmen Square. *See* China and political dissent
Toffler, A., 23, 146
Tonta, Y., and Kurbanoglu, S., 155
Touraine, Alain, 15–16, 36
  markets and democracy, 21
  on subjectivity, 32–34, 35
  success of groups and specific rights, 43
  vision of resisters, 35
TUBITAK, 155
Tun, Cohen
  email and Free Burma Movement, 75
Turgot, 24–25, 26
Turkey, 153–165
  education, employment and equality, 163–165
  effects of Internet on academia "Turkish University On-line" (project), 155–158

external pressures on Ministry of Education, 160
familiarity with foreign languages, 160–161
information technologies and employment, 164
Internet trends since "Turkish University On-line" project, 159–161
struggle for national identity, 154
technology and distance education (democraphic analysis), 159–160
women and cyberspace, 161–163
Turkish networks
Bitnet, 155
Computer Experimental Schools (CES) project, 159
cyberspace and sexual references, 162
democratization, 164–165
European Academic Research Network (EARN), 155
gendered roles in cyberspace, 163
informal questionnaire, 162–163
kadir (women's listserv), 161
"The Turkish University On-line," 155–159, 164
Turkish Universities and Research Institution Network (TUVAKA), 155
"Turkish University On-line" project
155-159, 164
absence of censorship, 156

beginnings and selected group, 155–156
freedom from cultural constraints, 157
new alliances and international coalitions, 157–158
social scientists and access to computers, 156, 157
students, 158
women, 156–157, 158
Turner, Alex, 80
development of Free Burma Web site, 81
social scientists and access to computers, 156, 157

U
Uchitel' skaia gazeta (Teacher' Gazette), 95–96, 98–99, 105
underground publications
Malawi, 51
United Front for Multiparty Democracy, 61
Urschel, J., 71
U.S. State Department advisory, 60

V
Veblen, 28
Velikhov, Evgenyi, 103, 104
Vygotsky, 100, 102
influence on educational reform, 99–100

W
Wajcman, J., 116, 121
Waters, R., 125
Watson, Thomas, 131
Weber, Max, 24, 28–29
attack on capitalism, 28

as contrasted to Karl Marx, 28
What is Democracy? (Touraine), 33
Wiebe, Robert H., 38
Williams, Raymond, 11
Williams, Rosalind, 24–26
women and cyberspace, 156–157, 158, 161–163
  gender and tradional roles, 162–163
women and telephone technology, 113–134
  contemporary status and research, 122–123
  a domestic technology, 119–122
  early phone company campaigns, 119
  expanding the counternarrative, 127–132
  gratification studies, 123–124
  historical relationship, 118–127
  importance of hotlines, 125–126
  kinkeeping, 124–125
  liberal feminism, 122–123, 127–128
  N. Frazer and public-sphere theory, 117–118
  perspective on narratives and counternarratives, 113–114
  phone and identity, 130–131
  phone as fetish, 128–129
  phone practices as instrumental or intrinsic, 123–126
  sex- or power-related interactions, 126–127, 128–129
  "sociability," 120
  solidarities and the formation of quasi-public spheres, 132–134
  tool of confinement/connection, 121–122
  use of "quasi-public" space, 113, 118, 125, 126, 127
World of Education, 105

Y
Yaobang, Hu, 141
Yeltsin, Boris, 148–149

Z
Zambia
  Malawi opposition politics, 61–62
  meeting in Lusaka, 61–62
Zapatista movement, 19–20
Zarni, 71, 169
Ziyang, Zhao, 141